S0-CWF-648

DATE DUE

UPI

Printed in USA

An Introduction to
COMPLEX ANALYSIS
IN SEVERAL VARIABLES

North-Holland Mathematical Library

Board of Advisory Editors:

M. Artin, H. Bass, J. Eells, W. Feit, P. J. Freyd, F. W. Gehring, H. Halberstam, L. Hörmander, M. Kac, J. H. B. Kemperman, H. A. Lauwerier, W. A. J. Luxemburg, P. F. Peterson, I. M. Singer and A. C. Zaanen

VOLUME 7

NORTH-HOLLAND PUBLISHING COMPANY – AMSTERDAM · LONDON
AMERICAN ELSEVIER PUBLISHING COMPANY, INC. – NEW YORK

An Introduction to COMPLEX ANALYSIS IN SEVERAL VARIABLES

LARS HÖRMANDER

Professor at the University of Lund

N·H
P·C

1973

NORTH-HOLLAND PUBLISHING COMPANY – AMSTERDAM · LONDON
AMERICAN ELSEVIER PUBLISHING COMPANY, INC. – NEW YORK

WITHDRAWN

ITHACA COLLEGE LIBRARY

© North-Holland Publishing Company – 1973

All Rights Reserved. No part of this publication may be reproduced, stored in a retrieval system, or transmitted, in any form or by any means, electronic, mechanical, photocopying, recording or otherwise, without the prior permission of the Copyright owner.

Library of Congress Catalog Card Number: 73-81532

North-Holland ISBN: *S* 0 7204 2450 X
0 7204 2457 7

American Elsevier ISBN: 0 444 10523 9

Published by:

North-Holland Publishing Company – Amsterdam
North-Holland Publishing Company, Ltd. – London

Sole distributors for the U.S.A. and Canada:

American Elsevier Publishing Company, Inc.
52 Vanderbilt Avenue
New York, N.Y. 10017

Printed in The Netherlands

PREFACE

Two recent developments in the theory of partial differential equations have caused this book to be written. One is the theory of overdetermined systems of differential equations with constant coefficients, which depends very heavily on the theory of functions of several complex variables. The other is the solution of the so-called $\bar{\partial}$ Neumann problem, which has made possible a new approach to complex analysis through methods from the theory of partial differential equations. Solving the Cousin problems with such methods gives automatically certain bounds for the solution, which are not easily obtained with the classical methods, and results of this type are important for the applications to overdetermined systems of differential equations. It has therefore seemed natural to give a self-contained exposition of complex analysis from the point of view of the theory of partial differential equations. Since we have concentrated on topics which are suitable for such a treatment, analytic spaces will not be discussed. Instead we have included some theorems on Banach algebras as another example of the applications to analysis of the theory of functions of several complex variables.

This book is only a slight modification of lecture notes from a course given by the author at Stanford University during the Spring and Summer quarters of 1964. The aim has not been to achieve completeness in any direction but to provide an easy introduction to complex analysis for readers whose main interest is in analysis. For this reason it has been assumed only that the reader knows a certain amount of real function theory, more specifically the elements of integration theory, distribution theory, functional analysis, and the calculus of differential forms. Very little algebra is used. In Chapter I the elementary theory of functions of a single complex variable is recalled briefly. The main reason for this is to introduce the central problems in a familiar case as a guide for the general case. Chapter I also includes some classical facts, such as the Cauchy integral formula for solutions of the inhomogeneous Cauchy-Riemann equations, which unfortunately are missing in many elementary texts. The last section of Chapter I develops the facts concerning subharmonic functions which are needed. Since most readers should pass quickly to Chapter II, we wish to mention that the main point of the Hartogs theorem on separate analyticity has been inserted there.

Chapter II starts with classical facts concerning power series expansions, domains of holomorphy, and pseudoconvex domains. Following a classical paper of Oka, rewritten in the spirit of differential equations, existence theorems for the Cauchy-Riemann equations in Runge domains are then proved. This is done to illustrate the Oka-Cartan methods in a very simple case which is sufficient for the main applications to the theory of Banach algebras. These are given in Chapter III where a preliminary section recalls the basic facts concerning such algebras. Both Chapter III and section 2.7 can be bypassed without any loss of the continuity.

In Chapter IV the Cauchy-Riemann equations are solved in domains of holomorphy by means of a variant of the $\bar{\partial}$ Neumann problem. At the same time a solution of the Levi problem is obtained, that is, the identity of pseudoconvex domains and domains of holomorphy is shown. These results are extended to Stein manifolds in Chapter V. It is proved that Stein manifolds can be embedded in complex vector spaces of high dimension. Chapter V ends with a proof that complex structures can be defined on a manifold by giving a system of Cauchy-Riemann equations satisfying a certain integrability condition.

Chapter VII is devoted to the theory of coherent analytic sheaves on Stein manifolds. The proofs are based on the existence theory for the Cauchy-Riemann equations established in Chapter V and the local theory presented in Chapter VI. A final section is devoted to "cohomology with bounds" for sheaves over $\mathbf{C}^n$ with polynomial generators. Used there are the existence theorems for the Cauchy-Riemann equations proved in Chapter IV. The book ends with applications to overdetermined systems of differential equations.

Preface

I am greatly indebted to colleagues and students at Stanford University who helped improve the original notes, and also to the National Science Foundation for supporting the work through grant GP 2426 at Stanford University during the summer of 1964.

Lars Hörmander

Princeton, New Jersey
January 1966

Preface to second edition

The main change in this edition is that section 4.4 has been improved. A number of references have also been added, particularly to work in the spirit of that section, and a few misprints have been corrected.

Lund in February 1973 Lars Hörmander

CONTENTS

LIST OF SYMBOLS

$\complement A$ is the complement of A (in some larger set understood from the context).

$\emptyset$ is the empty set.

$A \setminus B$ is a notation for $A \cap \complement B$.

$A \pm B = \{a \pm b; a \in A, b \in B\}$ if A and B are subsets of an abelian group.

$A \subset\subset B$ means that A is relatively compact in B, that is, A is contained in a compact subset of B.

∂A is the boundary of A.

$\partial_0 A$ denotes the distinguished boundary when A is a polydisc.

$C^k(\Omega)$, where Ω is an open set in $\mathbf{R}^N$ (or a C^∞ manifold) is the space of k times continuously differentiable complex valued functions in Ω, $0 \leq k \leq \infty$.

$C_0{}^k(A)$, where A is a subset of a C^∞ manifold Ω, denotes the set of functions in $C^k(\Omega)$ vanishing outside a compact subset of A.

supp f denotes the support of f, which is the closure of the smallest set outside which f vanishes (see p. 3).

D is sometimes used as a shorter notation for $C_0{}^\infty(\Omega)$ (see p. 78).

$A(\Omega)$ is the space of analytic functions in Ω (see pp. 1, 23).

$P(\Omega)$ is the space of plurisubharmonic functions in Ω (see p. 44).

$L^2(\Omega,\varphi)$ is the space of measurable functions in Ω such that (see pp. 78, 113)

$$\|u\|_\varphi^2 = \int |u|^2 e^{-\varphi}\, dx < \infty.$$

$\mathscr{D}'(\Omega)$ is the space of Schwartz distributions in Ω.

$\mathscr{E}'(\Omega)$ is the subspace of distributions with compact support.

W^s is the space of L^2 functions in $\mathbf{R}^N$ with all derivatives of order $\leq s$ in the sense of distribution theory belonging to L^2 (see p. 85).

$W^s(\Omega,\text{loc})$, where Ω is an open set in a C^∞ manifold, is the set of functions in Ω which agree on every compact subset of a coordinate patch with some function W^s in the coordinate space (see pp. 85, 119).

$L^2(\Omega,\text{loc})$ is the same as $W^0(\Omega,\text{loc})$.

$\mathscr{F}_{(p,q)}$, where $\mathscr{F}$ is any of the previous spaces, denotes the set of all forms of type (p,q) with coefficients in $\mathscr{F}$ (see p. 24).

List of Symbols

$\partial/\partial z_j$ and $\partial/\partial \bar{z}_j$ (see pp. 1 and 22).
$\partial^\alpha = (\partial/\partial z_1)^{\alpha_1} \cdots (\partial/\partial z_n)^{\alpha_n}$ (p. 26), where
α is a multiorder $= (\alpha_1, \cdots, \alpha_n)$ with α_j non-negative integers,
$|\alpha| = \alpha_1 + \cdots + \alpha_n$ and $\alpha! = \alpha_1! \cdots \alpha_n!$
$\wedge$ denotes exterior multiplication.
d is the exterior differentiation.
∂ and $\bar{\partial}$ are the components of d of type (1,0) and (0,1) (see pp. 22, 24).
u^*f, where f is a form and u a map, is defined on p. 23.
I (or J or K) often denotes a multi-index, that is, a sequence $(i_1, \cdots, i_p)$ of integers between 1 and n, the dimension of the space considered. We write $|I| = p$, and Σ'_I indicates that summation is restricted to multi-indices with $i_1 < i_2 < \cdots < i_p$.
$\hat{K}_\Omega$ is defined on pp. 8, 37, 109.
$\hat{K}_\Omega{}^P$ is defined on p. 46.
$\tilde{K}$ is defined on p. 53.
$\gamma_z f$ denotes the germ of f at z (see p. 152).
A_z denotes the set of germs at z of analytic functions.
D_T is the domain of the operator T.
R_T is the range of the operator T.
$d\lambda$ denotes the Lebesgue measure.
$\hat{f}$ denotes the Gelfand transform (or Fourier transform) of f.
$H^p(\mathscr{U}, \mathscr{F})$ is a cohomology group of the covering $\mathscr{U}$ with values in the sheaf $\mathscr{F}$ (see p. 177).
$H^p(X, \mathscr{F})$ is a cohomology group of the paracompact space X with values in the sheaf $\mathscr{F}$ (see p. 178).
$R[z]$ denotes the set of polynomials in one variable z with coefficients in the ring R.

Chapter I

ANALYTIC FUNCTIONS OF ONE COMPLEX VARIABLE

Summary. In the first two sections we recall the simplest properties of analytic functions which follow from the Cauchy integral formula. Then follows a discussion of approximation theorems (the Runge theorem) and existence theorems for meromorphic functions (the Mittag-Leffler and Weierstrass theorems). These are the one-dimensional case of the Cousin problems around which the theory of analytic functions of several variables has developed. Finally we prove some basic theorems concerning subharmonic functions.

1.1. Preliminaries. Let u be a complex valued function in $C^1(\Omega)$,† where Ω is an open set in the complex plane $\mathbf{C}$, which we identify with $\mathbf{R}^2$. If the real coordinates are denoted by x, y, and $z = x + iy$, we have $2x = z + \bar{z}$, $2iy = z - \bar{z}$, so that the differential of u can be expressed as a linear combination of dz and $d\bar{z}$,

$$du = \frac{\partial u}{\partial x}\,dx + \frac{\partial u}{\partial y}\,dy = \frac{\partial u}{\partial z}\,dz + \frac{\partial u}{\partial \bar{z}}\,d\bar{z}, \tag{1.1.1}$$

where we have used the notations

$$\frac{\partial u}{\partial z} = \frac{1}{2}\left(\frac{\partial u}{\partial x} + \frac{1}{i}\frac{\partial u}{\partial y}\right), \qquad \frac{\partial u}{\partial \bar{z}} = \frac{1}{2}\left(\frac{\partial u}{\partial x} - \frac{1}{i}\frac{\partial u}{\partial y}\right). \tag{1.1.2}$$

Definition 1.1.1. *A function $u \in C^1(\Omega)$ is said to be analytic (or holomorphic) in Ω if $\partial u/\partial \bar{z} = 0$ in Ω (the Cauchy–Riemann equation), or equivalently if du is proportional to dz. For analytic functions one also writes u' instead of $\partial u/\partial z$; thus $du = u'\,dz$ if u is analytic. The set of all analytic functions in Ω is denoted by $A(\Omega)$.*

† For the notation used in this book not otherwise explained, see list of symbols on p. ix.

Examples. (1) For every integer n we have $d(z^n) = nz^{n-1}\,dz$ (for $z \neq 0$ if $n < 0$). Hence every polynomial $p(z) = \Sigma_0^n a_k z^k$ is an analytic function, and $p'(z) = \Sigma_1^n k a_k z^{k-1}$. (2) If we define $e^z = e^x(\cos y + i \sin y)$, we obtain $d\,e^z = e^z\,dz$ so e^z is analytic.

Since the differential operator $\partial/\partial\bar{z}$ is linear, it is obvious that linear combinations with complex coefficients of analytic functions are analytic. From the product rule $d(uv) = u\,dv + v\,du$ we obtain the product rule for the operators $\partial/\partial z$ and $\partial/\partial\bar{z}$. Hence the product of analytic functions is analytic.

Let u be analytic in Ω and let v be analytic in (an open set containing) the range of u. Then the function $z \to v(u(z))$ is analytic in Ω, for the chain rule gives

$$dv = v'(u)\,du = v'(u)u'(z)\,dz,$$

which also implies that $\partial v/\partial z = (\partial v/\partial u)(\partial u/\partial z)$.

We shall finally study the inverse of an analytic function. First note that since $du = u'\,dz$, the map $dz \to du$ is a rotation followed by a dilation in the ratio $|u'|$. Hence the Jacobian of the map $z \to u(z)$, considered as a map of $\mathbf{R}^2$ into $\mathbf{R}^2$, is equal to $|u'|^2$. If $u'(z_0) \neq 0$, it follows therefore from the implicit function theorem that u maps a neighborhood of z_0 homeomorphically on a neighborhood of $u_0 = u(z_0)$, and that the inverse map $u \to z(u)$ is also continuously differentiable in a neighborhood of u_0. Since $u(z(w)) = w$, the chain rule gives $u'(z(w))\,dz = dw$, so z is an analytic function of w and $\partial z(w)/\partial w = 1/u'(z(w))$.

1.2. Cauchy's integral formula and its applications. Let ω be a bounded open set in $\mathbf{C}$, such that the boundary $\partial\omega$ consists of a finite number of C^1 Jordan curves. Stokes' formula gives, if $u \in C^1(\bar{\omega})$,

$$\int_{\partial\omega} u\,dz = \iint_\omega du \wedge dz, \tag{1.2.1}$$

or if we note that $du \wedge dz = \partial u/\partial\bar{z}\,d\bar{z} \wedge dz = 2i\,\partial u/\partial\bar{z}\,dx \wedge dy$

$$\int_{\partial\omega} u\,dz = 2i\iint_\omega \partial u/\partial\bar{z}\,dx \wedge dy = \iint_\omega \partial u/\partial\bar{z}\,d\bar{z} \wedge dz. \tag{1.2.2}$$

(This can of course be proved directly by integrating the right-hand side.) Here $\partial\omega$ is oriented so that ω lies to the left of $\partial\omega$. An immediate consequence is that $\int_{\partial\omega} u\,dz = 0$ if $u \in C^1(\bar{\omega})$ and u is analytic in ω. Moreover, we obtain Cauchy's integral formula:

Theorem 1.2.1. *If* $u \in C^1(\bar{\omega})$, *we have*

$$(1.2.3)\quad u(\zeta) = (2\pi i)^{-1}\left\{\int_{\partial\omega} \frac{u(z)}{z-\zeta}\,dz + \iint_{\omega} \frac{\partial u/\partial \bar{z}}{z-\zeta}\,dz \wedge d\bar{z}\right\}, \qquad \zeta \in \omega.$$

Proof. Put $\omega_\varepsilon = \{z; z \in \omega, |z-\zeta| > \varepsilon\}$ where $0 < \varepsilon <$ the distance from ζ to $\complement\omega$. If we apply (1.2.2) to $u(z)/(z-\zeta)$ and note that $1/(z-\zeta)$ is analytic in ω_ε, we obtain

$$\iint_{\omega_\varepsilon} \partial u/\partial\bar{z}\,(z-\zeta)^{-1}\,d\bar{z} \wedge dz = \int_{\partial\omega} u(z)(z-\zeta)^{-1}\,dz - \int_0^{2\pi} u(\zeta + \varepsilon e^{i\theta})i\,d\theta.$$

Since $(z-\zeta)^{-1}$ is integrable over ω and u is continuous at ζ, we obtain (1.2.3) by letting $\varepsilon \to 0$.

Conversely, we shall prove

Theorem 1.2.2. *If* μ *is a measure with compact support*† *in* $\mathbf{C}$, *the integral*

$$u(\zeta) = \int (z-\zeta)^{-1}\,d\mu(z)$$

defines an analytic C^∞ *function outside the support of* μ. *In any open set* ω *where* $d\mu = (2\pi i)^{-1}\varphi\,dz \wedge d\bar{z}$ *for some* $\varphi \in C^k(\omega)$, *we have* $u \in C^k(\omega)$ *and* $\partial u/\partial\bar{z} = \varphi$ *if* $k \geq 1$.

Proof. That $u \in C^\infty$ outside the support K of μ is obvious since $(z-\zeta)^{-1}$ is a C^∞ function of (z,ζ) when $z \in K$ and $\zeta \in \complement K$, and since $\partial(z-\zeta)^{-1}/\partial\bar{\zeta} = 0$ when $\zeta \neq z$, the analyticity follows by differentiation under the sign of integration. To prove the second statement we first assume that $\omega = \mathbf{R}^2$. After a change of variables we can write

$$u(\zeta) = -(2\pi i)^{-1}\iint \varphi(\zeta - z)z^{-1}\,dz \wedge d\bar{z}.$$

Since z^{-1} is integrable on every compact set, it is legitimate to differentiate under the sign of integration at most k times and the integrals obtained are continuous. Hence $u \in C^k$ and

$$\begin{aligned}\partial u/\partial\bar{\zeta} &= -(2\pi i)^{-1}\iint \partial\varphi(\zeta - z)/\partial\bar{\zeta}\, z^{-1}\,dz \wedge d\bar{z}\\ &= (2\pi i)^{-1}\iint (z-\zeta)^{-1}\,\partial\varphi(z)/\partial\bar{z}\,dz \wedge d\bar{z}.\end{aligned}$$

Application of Theorem 1.2.1 with u replaced by φ and ω equal to a disc containing the support of φ now gives $\partial u/\partial\bar{\zeta} = \varphi$. Finally, if ω is arbitrary, we can, for every $z_0 \in \omega$, choose a function $\psi \in C_0{}^k(\omega)$ which

† The support of a measure or function is the smallest closed set outside which it is equal to 0.

is equal to 1 in a neighborhood V of z_0. If $\mu_1 = \psi\mu$ and $\mu_2 = (1 - \psi)\mu$, we have $u = u_1 + u_2$ where

$$u_j(\zeta) = \int (z - \zeta)^{-1}\, d\mu_j(\zeta).$$

Since μ_1 is equal to $(2\pi i)^{-1}\psi\varphi\, dz \wedge d\bar{z}$ and $\psi\varphi \in C_0{}^k(\mathbf{R}^2)$, we have $u_1 \in C^k$ and $\partial u_1/\partial\bar{\zeta} = \psi\varphi$. Since μ_2 vanishes in V, it follows that $u \in C^k(V)$ and that $\partial u/\partial\bar{\zeta} = \varphi$ in V. The proof is complete.

Corollary 1.2.3. *Every* $u \in A(\Omega)$ *is in* $C^\infty(\Omega)$. *Hence* $u' \in A(\Omega)$ *if* $u \in A(\Omega)$.

Proof. This follows from Theorems 1.2.1 and 1.2.2 applied to discs ω with $\bar{\omega} \subset \Omega$.

More precise information is given in the next theorem.

Theorem 1.2.4. *For every compact set* $K \subset \Omega$ *and every open neighborhood* $\omega \subset \Omega$ *of* K *there are constants* C_j, $j = 0, 1, \cdots$, *such that*

$$\sup_{z \in K} |u^{(j)}(z)| \leq C_j \|u\|_{L^1(\omega)}, \qquad u \in A(\Omega), \tag{1.2.4}$$

where $u^{(j)} = \partial^j u/\partial z^j$.

Proof. Choose $\psi \in C_0{}^\infty(\omega)$ so that $\psi = 1$ in a neighborhood of K. If $u \in A(\Omega)$, we have $\partial(\psi u)/\partial\bar{z} = u\partial\psi/\partial\bar{z}$ and consequently Theorem 1.2.1 applied to ψu gives

$$\psi(\zeta)u(\zeta) = (2\pi i)^{-1} \int u(z)\, \partial\psi/\partial\bar{z}\, (z - \zeta)^{-1}\, dz \wedge d\bar{z}. \tag{1.2.5}$$

Since $\psi = 1$ in a neighborhood of K and $|z - \zeta|$ is bounded from below when $\zeta \in K$ and z is in the support of $\partial\psi/\partial\bar{z}$, differentiation of (1.2.5) leads immediately to (1.2.4).

Corollary 1.2.5. *If* $u_n \in A(\Omega)$ *and* $u_n \to u$ *when* $n \to \infty$, *uniformly on compact subsets of* Ω, *it follows that* $u \in A(\Omega)$.

Proof. Application of (1.2.4) to $u_n - u_m$ shows that $\partial u_n/\partial z$ converges uniformly. Since $\partial u_n/\partial\bar{z} = 0$, it follows that $\partial u_n/\partial x$ and $\partial u_n/\partial y$ converge uniformly on compact sets. Hence $u \in C^1$ and $\partial u/\partial\bar{z} = \lim \partial u_n/\partial\bar{z} = 0$.

Corollary 1.2.6. (*Stieltjes-Vitali*) *If* $u_n \in A(\Omega)$ *and the sequence* $|u_n|$ *is uniformly bounded on every compact subset of* Ω, *there is a subsequence* u_{n_j} *converging uniformly on every compact subset of* Ω *to a limit* $u \in A(\Omega)$.

Proof. As in Corollary 1.2.5, we obtain from Theorem 1.2.4 that there are uniform bounds for the first-order derivatives of u_n on any compact set. Hence this sequence is equicontinuous and the corollary follows from Ascoli's theorem and Corollary 1.2.5.

Corollary 1.2.7. *The sum of a power series*

$$u(z) = \sum_0^\infty a_n z^n$$

is analytic in the interior of the circle of convergence.

Proof. The series converges uniformly in every smaller disc.

Theorem 1.2.8. *If u is analytic in $\Omega = \{z;\ |z| < r\}$, we have*

$$u(z) = \sum_0^\infty u^{(n)}(0)\, z^n/n!$$

with uniform convergence on every compact subset of Ω.

Proof. Let $r_1 < r_2 < r$. We have by (1.2.3)

$$u(z) = (2\pi i)^{-1} \int_{|\zeta| = r_2} u(\zeta)/(\zeta - z)\, d\zeta, \qquad |z| \leq r_1. \tag{1.2.6}$$

Since

$$(\zeta - z)^{-1} = \sum_0^\infty z^n \zeta^{-n-1}, \qquad |z| \leq r_1, \qquad |\zeta| = r_2,$$

and the series is uniformly and absolutely convergent, the theorem follows if we integrate term by term, noting that (1.2.6) gives

$$u^{(n)}(0) = n!(2\pi i)^{-1} \int_{|\zeta| = r_2} u(\zeta)\zeta^{-n-1}\, d\zeta.$$

Corollary 1.2.9. (*The uniqueness of analytic continuation.*) *If $u \in A(\Omega)$ and there is some point z in Ω where*

$$u^{(k)}(z) = 0, \quad \textit{for all } k \geq 0, \tag{1.2.7}$$

it follows that $u = 0$ in Ω if Ω is connected.

Proof. The set of all $z \in \Omega$ satisfying (1.2.7) is obviously closed in Ω, and by Theorem 1.2.8 it is also open. Since it is non-empty by assumption, it must be equal to Ω.

Corollary 1.2.10. *If u is analytic in the disc $\Omega = \{z;\ |z| < r\}$ and if u is not identically* 0, *one can write u in one and only one way in the form*

$$u(z) = z^n v(z)$$

where n is an integer ≥ 0 *and* $v \in A(\Omega)$, $v(0) \neq 0$ *(which means that* $1/v$ *is also analytic in a neighborhood of* 0*).*

Proof. The proof is obvious.

Theorem 1.2.11. *If u is analytic in* $\{z;\ |z - z_0| < r\} = \Omega$ *and if* $|u(z)| \leq |u(z_0)|$ *when* $z \in \Omega$, *then u is constant in* Ω.

Proof. We may assume that $u(z_0) \neq 0$. Since

$$u(z_0) = (2\pi)^{-1} \int_0^{2\pi} u(z_0 + \rho e^{i\theta})\, d\theta$$

when $0 < \rho < r$, we obtain

$$\int_0^{2\pi} (1 - u(z_0 + \rho e^{i\theta})/u(z_0))\, d\theta = 0.$$

The real part of the integrand is ≥ 0 and $= 0$ only when $u(z_0) = u(z_0 + \rho e^{i\theta})$. This proves the theorem.

Corollary 1.2.12. *(Maximum principle.) Let* Ω *be bounded and let* $u \in C(\bar{\Omega})$ *be analytic in* Ω. *Then the maximum of* $|u|$ *in* $\bar{\Omega}$ *is attained on the boundary.*

Proof. If the maximum is attained in an interior point, Theorem 1.2.11 and Corollary 1.2.9 prove that u is constant in the component of Ω containing that point and therefore $|u|$ assumes the same value at some boundary point.

1.3. The Runge approximation theorem. From Theorem 1.2.8 it follows in particular that a function which is analytic in a disc can be approximated uniformly by polynomials in z on any smaller disc. In particular, every entire function can be approximated by polynomials uniformly on every compact set. We shall now give a general approximation theorem.

Theorem 1.3.1. *(Runge.) Let* Ω *be an open set in* $\mathbf{C}$ *and K a compact subset of* Ω. *The following conditions on* Ω *and on K are equivalent:*

(a) *Every function which is analytic in a neighborhood of K can be approximated uniformly on K by functions in* $A(\Omega)$.

(b) *The open set* $\Omega \setminus K = \Omega \cap \complement K$ *has no component which is relatively compact in* Ω.

(c) *For every* $z \in \Omega \setminus K$ *there is a function* $f \in A(\Omega)$ *such that*

$$|f(z)| > \sup_K |f|. \tag{1.3.1}$$

By the remarks preceding the theorem we obtain the following special case by taking $\Omega = \mathbf{C}$.

Corollary 1.3.2. *Every function which is analytic in a neighborhood of the compact set K can be approximated by polynomials uniformly on K if and only if* $\complement K$ *is connected, or equivalently, for every* $z \in \complement K$ *there is a polynomial f such that* (1.3.1) *is valid.*

Proof of Theorem 1.3.1. We first prove that (c) $\Rightarrow$ (b) and that (a) $\Rightarrow$ (b). Thus assume that (b) is not valid, that is, that $\Omega \setminus K$ has a component O such that $\bar{O}$ is compact and $\subset \Omega$. Then the boundary of O is a subset of K and the maximum principle gives

$$\sup_{\bar{O}} |f| \leq \sup_{K} |f|, \qquad f \in A(\Omega), \tag{1.3.2}$$

which contradicts (c). If (a) were valid we could for every f which is analytic in a neighborhood of K choose $f_n \in A(\Omega)$ so that $f_n \to f$ uniformly on K. Application of (1.3.2) to $f_n - f_m$ proves that f_n converges uniformly in $\bar{O}$ to a limit F. We have $F = f$ on the boundary of O, and F is analytic in O and continuous in $\bar{O}$. In particular, we can choose $f(z) = 1/(z - \zeta)$ if $\zeta \in O$, and then we have $(z - \zeta)F(z) = 1$ on the boundary of O, hence $(z - \zeta)F(z) = 1$ in O. This gives a contradiction when $z = \zeta$.

To prove that (b) $\Rightarrow$ (a) it suffices to show that every measure μ on K which is orthogonal to $A(\Omega)$ is also orthogonal to every function f which is analytic in a neighborhood of K, for the theorem is then a consequence of the Hahn–Banach theorem. Set

$$\varphi(\zeta) = \int (z - \zeta)^{-1}\, d\mu(z), \qquad \zeta \in \complement K.$$

By Theorem 1.2.2, φ is analytic in $\complement K$, and when $\zeta \in \complement \Omega$ we have

$$\varphi^{(k)}(\zeta) = k! \int (z - \zeta)^{-k-1}\, d\mu(z) = 0 \quad \text{for every } k,$$

for the function $z \to (z - \zeta)^{-k-1}$ is analytic in Ω if $\zeta \in \complement \Omega$. Hence $\varphi = 0$ in every component of $\complement K$ which intersects $\complement \Omega$. Since $\int z^n\, d\mu(z) = 0$ for every n and $(z - \zeta)^{-1}$ can be expanded in a power series in z which converges uniformly on K if $|\zeta| > \sup_K |z|$, we also have $\varphi = 0$ in the unbounded component of $\complement K$. Now (b) guarantees that $\Omega \setminus K$ has no component which is relatively compact in Ω, and we conclude that $\varphi = 0$ in $\complement K$.

Choose a function $\psi \in C_0^\infty(\omega)$, where ω is a neighborhood of K in which f is analytic, and choose ψ so that $\psi = 1$ on K. Then we have

$$f(z) = \psi(z) f(z) = (2\pi i)^{-1} \iint f(\zeta)\, \partial\psi(\zeta)/\partial\bar{\zeta}\, (\zeta - z)^{-1}\, d\zeta \wedge d\bar{\zeta}, \qquad z \in K.$$

Since $\partial\psi/\partial\bar{\zeta} = 0$ in a neighborhood of K, inverting the order of integrations gives

$$\int f(z)\, d\mu(z) = -(2\pi i)^{-1} \iint f(\zeta)\, \partial\psi(\zeta)/\partial\bar{\zeta}\ \varphi(\zeta)\, d\zeta \wedge d\bar{\zeta} = 0.$$

Hence f can be approximated on K by functions in $A(\Omega)$, which proves the equivalence of (a) and (b).

Finally, to prove that (b) $\Rightarrow$ (c), we assume that (b) is fulfilled and let $z \in \Omega \setminus K$. Choose a closed disc L with center at z so that $L \subset \Omega \setminus K$. Then the components of $\Omega \setminus (K \cup L)$ are the same as those of $\Omega \setminus K$ apart from the fact that L has been removed from one of the components. Hence $K \cup L$ also satisfies (b). According to (a) the function which is 0 in a neighborhood of K and 1 in a neighborhood of L can therefore be approximated uniformly by functions in $A(\Omega)$. Hence we can find $f \in A(\Omega)$ so that

$$|f| < 1/2 \quad \text{in } K, \ |f - 1| < 1/2 \quad \text{in } L.$$

This proves (c).

If K is an arbitrary compact subset of Ω, we define the $A(\Omega)$-hull $\hat{K}$ of K by

$$\hat{K} = \hat{K}_\Omega = \{z;\, z \in \Omega,\, |f(z)| \le \sup_K |f| \quad \text{for every } f \in A(\Omega)\}.$$

If we choose $f(z) = 1/(z - \zeta)$ where $\zeta \in \complement\Omega$, we obtain

$$d(K, \complement\Omega) = d(\hat{K}, \complement\Omega),$$

where d denotes the distance, and if we consider $f(z) = e^{az}$ for every complex number a, we obtain

$$\hat{K} \subset \text{convex hull of } K.$$

Furthermore, it is clear that $\hat{\hat{K}} = \hat{K}$. For every compact set $K \subset \Omega$ the hull $\hat{K}$ is thus a compact subset of Ω containing K for which the hypotheses of the Runge approximation theorem are fulfilled. One can therefore choose an increasing sequence K_j of compact subsets of Ω such that $K_j = \hat{K}_j$ and every compact subset of Ω belongs to K_j for some j.

We can also give a description of $\hat{K}$ analogous to condition (b) in Theorem 1.3.1.

Theorem 1.3.3. *$\hat{K}_\Omega$ is the union of K and the components of $\Omega \setminus K$ which are relatively compact in Ω.*

Proof. If O is a component of $\Omega \setminus K$ which is relatively compact in Ω, we have the inequality (1.3.2) so that $O \subset \hat{K}$. The union K_1 of K and all such components is therefore contained in $\hat{K}$. Now $\Omega \setminus K_1$ is open since it is a union of open components of $\Omega \setminus K$. Hence K_1 is compact, and by the definition of K_1 no component of $\Omega \setminus K_1$ is relatively compact in Ω. Thus K_1 satisfies condition (b) in Theorem 1.3.1, so that condition (c) gives $K_1 = \hat{K}_1 \supset \hat{K}$. The proof is complete.

We shall now give a variant of the Runge theorem for two open sets.

Theorem 1.3.4. *Let $\Omega_1 \subset \Omega_2$ be open sets in* $\mathbf{C}$. *The following conditions are equivalent:*

(a) *Every function in $A(\Omega_1)$ can be approximated by functions in $A(\Omega_2)$, uniformly on every compact subset of Ω_1.*

(b) *If $\Omega_2 \setminus \Omega_1 = L \cup F$ where F is closed in Ω_2 and L is compact, $F \cap L = \emptyset$, it follows that L is empty.*

(c_1) *For every compact set $K \subset \Omega_1$ we have $\hat{K}_{\Omega_2} = \hat{K}_{\Omega_1}$.*

(c_2) *For every compact set $K \subset \Omega_1$ we have $\hat{K}_{\Omega_2} \cap \Omega_1 = \hat{K}_{\Omega_1}$.*

(c_3) *For every compact set $K \subset \Omega_1$ the set $\hat{K}_{\Omega_2} \cap \Omega_1$ is compact.*

Proof. It is obvious that (a) $\Rightarrow$ (c_2) $\Rightarrow$ (c_3). If we set $K' = \hat{K}_{\Omega_2} \cap \Omega_1$ and $K'' = \hat{K}_{\Omega_2} \cap \complement \Omega_1$, it follows from ($c_3$) that the disjoint sets K' and K'' are compact. If $f \in A(\Omega_1)$, it follows from Theorem 1.3.1 that the function which is equal to f on K' and equal to 1 on K'' can be approximated uniformly on $K' \cup K''$ by functions in $A(\Omega_2)$. This proves (a), and choosing $f = 0$ we obtain $K'' = \emptyset$, hence $\hat{K}_{\Omega_2} = \hat{K}_{\Omega_2} \cap \Omega_1$. Since (a) $\Rightarrow$ (c_2), we conclude that (c_1) is valid, so (a), (c_1), (c_2), (c_3) are equivalent.

To prove that (c_1) $\Rightarrow$ (b), we choose an open set $\omega \supset L$ so that $\bar{\omega} \cap F$ is empty and ω is relatively compact in Ω_2. Since $\partial\omega \cap (\Omega_2 \setminus \Omega_1) = \emptyset$ and $\partial\omega \subset \Omega_2$, we have $\partial\omega \subset \Omega_1$, and by the maximum principle the $A(\Omega_2)$-hull of $\partial\omega$ contains ω and therefore L. Hence $L \subset \Omega_1$ in view of (c_1), so that $L = \emptyset$.

To prove that (b) $\Rightarrow$ (c_1), we consider a component O of $\Omega_2 \setminus K$ which is relatively compact in Ω_2. Since $\partial O \subset K \subset \Omega_1$, the set

$$L = \bar{O} \cap (\Omega_2 \setminus \Omega_1) = O \cap (\Omega_2 \setminus \Omega_1)$$

is a compact subset of O, and since $(\complement O) \cap (\Omega_2 \setminus \Omega_1)$ is closed in Ω_2, it follows from (b) that $L = \emptyset$. Hence $O \subset \Omega_1$, so it follows from Theorem 1.3.3 that $\hat{K}_{\Omega_2} \subset \hat{K}_{\Omega_1}$. Since the opposite inclusion is obvious, condition (c_1) follows.

1.4. The Mittag-Leffler theorem. We shall first give a definition of meromorphic functions which is somewhat complicated but has the advantage that it can be used in the case of several variables.

For every $z \in \mathbf{C}$ we let A_z be the set of equivalence classes of functions f which are analytic in some neighborhood of z, with the equivalence relation $f \sim g$ if $f = g$ in a neighborhood of z. If f is analytic in a neighborhood of z, we write f_z for the residue class of f in A_z. It is clear that A_z is a ring without divisors of 0 so we can form the quotient field M_z of A_z.

Definition 1.4.1. *A meromorphic function f in the open set $\Omega \subset \mathbf{C}$ is a map*

$$\varphi : \Omega \to \bigcup_z M_z$$

such that $\varphi(z) \in M_z$ for every z and to every point in Ω there is a neighborhood ω and functions $f, g \in A(\omega)$ such that $\varphi(z) = f_z/g_z$ when $z \in \omega$. The set of all meromorphic functions in Ω is denoted by $M(\Omega)$.

In particular, if $F \in A(\Omega)$, the map $z \to F_z$ is a meromorphic function, and since different analytic functions define different meromorphic functions, we can identify $A(\Omega)$ with a subset of $M(\Omega)$. It is convenient to use the notation φ_z instead of $\varphi(z)$ also for an arbitrary meromorphic function φ.

The meromorphic functions form a ring where each element which does not vanish identically in any component of Ω has an inverse.

To every $q \in M_\zeta$ we can assign a value $q(\zeta)$ at ζ. To do so we choose f and g analytic in a neighborhood of ζ so that $q = f_\zeta/g_\zeta$; thus $g_\zeta \neq 0$. If $q = 0$ we set $q(\zeta) = 0$. When $q \neq 0$, it follows from Corollary 1.2.10 that we can write $f(z) = (z - \zeta)^n f_1(z)$ and $g(z) = (z - \zeta)^m g_1(z)$ where $f_1(\zeta) g_1(\zeta) \neq 0$ and f_1, g_1 are analytic in a neighborhood of ζ. It is clear that $n - m$ and $f_1(\zeta)/g_1(\zeta)$ only depend on q and not on the choice of f and g. Hence we can define

$$q(\zeta) = \begin{cases} \infty & \text{if } n < m \\ f_1(\zeta)/g_1(\zeta) & \text{if } n = m \\ 0 & \text{if } n > m. \end{cases}$$

If $\varphi \in M(\Omega)$, we obtain a map

$$z \to \varphi_z(z) = F(z) \in \mathbf{C} \cup \{\infty\}$$

such that F is analytic in the complement of a discrete subset D of Ω and $1/F$ is analytic in a neighborhood of D (we set $1/\infty = 0$). Conversely, if we have a function F with these properties, a meromorphic function φ is defined by

$$\varphi_z = F_z \quad \text{if } z \notin D, \qquad \varphi_z = 1/(1/F)_z \quad \text{if } z \in D,$$

and $\varphi_z(z) = F(z)$ for every z. Hence Definition 1.4.1 is equivalent to the classical definition of a meromorphic function, for the correspondence between F and φ which we have given is one-to-one. The points with $F(z) = \infty$ are the poles of F. In what follows we do not distinguish between φ and F.

Theorem 1.4.2. *If F is meromorphic in a neighborhood of ζ, then there is a neighborhood of ζ, where*

$$F(z) = \sum_1^n A_k(z - \zeta)^{-k} + G(z)$$

with constants A_k and an analytic function G. The representation is unique. If $F_\zeta \neq 0$ there is also a unique representation of the form

$$F(z) = (z - \zeta)^n G(z),$$

where $G(\zeta) \neq 0$ and n is an integer. If $n > 0$, we have a zero of order n at ζ; and if $n < 0$, we have a pole of order $-n$.

The proof is an obvious consequence of Corollary 1.2.10. We shall next discuss the first representation; the multiplicative representation will be studied in section 1.5.

Theorem 1.4.3. (*Mittag-Leffler*) *Let z_j, $j = 1, 2, \cdots$, be a discrete sequence of different points in the open set Ω, and let f_j be meromorphic in a neighborhood of z_j. Then there exists a meromorphic function f in Ω such that f is analytic outside the points z_j and $f - f_j$ is analytic in a neighborhood of z_j for every j.*

Proof. In view of Theorem 1.4.2 we may assume that

$$f_j(z) = \sum_1^{n_j} A_{jk}(z - z_j)^{-k}.$$

We want to find functions $u_j \in A(\Omega)$ so that the series

$$f(z) = \sum_1^\infty (f_j(z) - u_j(z))$$

defines a function f with the required properties. To do so we choose an increasing sequence of compact sets $K_j \subset \Omega$ with $\hat{K}_j = K_j$ so that every compact subset of Ω is contained in some K_j. We may assume that $z_k \notin K_j$ when $k \geq j$, since the points z_k have no accumulation point in Ω. By the Runge theorem we can then choose $u_j \in A(\Omega)$ so that

$$|f_j(z) - u_j(z)| < 2^{-j}$$

in K_j. But then the series

$$\sum_{k}^{\infty} (f_j(z) - u_j(z))$$

converges uniformly on K_k to a function which is analytic in the interior of K_k. Hence the definition of f above is meaningful and f has the required properties.

Another formulation of the Mittag-Leffler theorem is the following, which must be used in the case of several variables:

Theorem 1.4.3′. *Let $\Omega = \cup_j \Omega_j$ where Ω_j are open sets in $\mathbf{C}$. If $f_j \in M(\Omega_j)$ and $f_j - f_k \in A(\Omega_j \cap \Omega_k)$ for all j and k, one can find $f \in M(\Omega)$ so that $f - f_j \in A(\Omega_j)$ for every j.*

That this is equivalent to Theorem 1.4.3 may be verified by the reader. Another equivalent result is the following:

Theorem 1.4.4. *For every $f \in C^\infty(\Omega)$ the equation $\partial u/\partial \bar{z} = f$ has a solution $u \in C^\infty(\Omega)$.*

Proof. Choose an increasing sequence of compact sets $K_j \subset \Omega$ with $\hat{K}_j = K_j$ so that every compact subset of Ω is contained in some K_j. Let $\psi_j \in C_0^\infty(\Omega)$ be equal to 1 in a neighborhood of K_j and set $\varphi_1 = \psi_1$, $\varphi_j = \psi_j - \psi_{j-1}$ when $j > 1$. Then $\varphi_j = 0$ in a neighborhood of K_{j-1} and $\Sigma_1^\infty \varphi_j = 1$ in Ω. By Theorem 1.2.2 we can find $u_j \in C^\infty(\mathbf{R}^2)$ so that $\partial u_j/\partial \bar{z} = \varphi_j f$. This means in particular that u_j is analytic in a neighborhood of K_{j-1}. By the Runge theorem we can therefore choose $v_j \in A(\Omega)$ so that $|u_j - v_j| < 2^{-j}$ in K_{j-1}. Then the sum

$$u = \sum_{1}^{\infty} (u_j - v_j)$$

is uniformly convergent on every compact set in Ω. The sum from $l + 1$ to ∞ consists of terms which are analytic near K_l and it converges uniformly on K_l to a function which is analytic in the interior of K_l. Hence $u \in C^\infty(\Omega)$, and since $\partial u/\partial \bar{z}$ can be computed by termwise differentiation, we have

$$\partial u/\partial \bar{z} = \sum_{1}^{\infty} \varphi_j f = f.$$

This completes the proof.

We shall now show that Theorem 1.4.4 implies a strengthened form of Theorem 1.4.3′.

Theorem 1.4.5. *Let* $\Omega = \cup_1^\infty \Omega_j$ *and let* $g_{jk} \in A(\Omega_j \cap \Omega_k)$, $j, k = 1, 2, \cdots$ *satisfy the conditions*

(1.4.1) $$g_{jk} = -g_{kj}, \quad g_{jk} + g_{kl} + g_{lj} = 0 \quad \text{in } \Omega_j \cap \Omega_k \cap \Omega_l \text{ for all } j, k, l.$$

Then one can find $g_j \in A(\Omega_j)$ *so that*

(1.4.2) $$g_{jk} = g_k - g_j \quad \text{in } \Omega_j \cap \Omega_k \text{ for all } j \text{ and } k.$$

Proof that Theorem 1.4.5 implies Theorem 1.4.3′. With the notations of Theorem 1.4.3′ we set $g_{jk} = f_j - f_k$. The hypothesis (1.4.1) of Theorem 1.4.5 is then fulfilled, so we can find $g_j \in A(\Omega_j)$ satisfying the equations

$$f_j - f_k = g_{jk} = g_k - g_j \quad \text{in } \Omega_j \cap \Omega_k \text{ for all } j \text{ and } k.$$

This means that $f_j + g_j = f_k + g_k$ in $\Omega_j \cap \Omega_k$. Hence there is a meromorphic function f in Ω such that $f = f_j + g_j$ in Ω_j for every j and, since $f - f_j = g_j \in A(\Omega_j)$, this proves Theorem 1.4.3′.

Proof of Theorem 1.4.5. We can choose a partition of unity subordinate to the covering $\{\Omega_j\}$, that is, we can choose functions φ_ν and positive integers i_ν, $\nu = 1, 2, \cdots$ so that

(i) $$\varphi_\nu \in C_0^\infty(\Omega_{i_\nu}).$$

(ii) All but a finite number of functions φ_ν vanish identically on any compact subset of Ω.

(iii) $$\sum \varphi_\nu = 1 \quad \text{on } \Omega.$$

(See for example Schwartz [1] or de Rham [1].)

If (1.4.2) is fulfilled, we obtain by taking $j = i_\nu$, multiplying by φ_ν and adding

$$g_k = h_k + u,$$

where we have written

$$h_k = \sum \varphi_\nu g_{i_\nu k}$$

and $u = \Sigma\, \varphi_\nu g_{i_\nu}$. The function h_k is well defined in terms of the functions g_{jk} (we set $\varphi_\nu g_{i_\nu k} = 0$ outside Ω_{i_ν}) and belongs to $C^\infty(\Omega_k)$. Furthermore, we have

$$h_k - h_j = \sum \varphi_\nu (g_{i_\nu k} - g_{i_\nu j}) = \sum \varphi_\nu g_{jk} = g_{jk}$$

since $g_{ki_\nu} + g_{i_\nu j} + g_{jk} = 0$. This implies that

$$\partial h_k / \partial \bar{z} = \partial h_j / \partial \bar{z} \quad \text{in } \Omega_j \cap \Omega_k.$$

Hence there is a function $\psi \in C^\infty(\Omega)$ such that

$$\psi = \partial h_k/\partial \bar{z}, \quad \text{in } \Omega_k \text{ for every } k.$$

If we now choose u as a solution of the equation

$$\partial u/\partial \bar{z} = -\psi,$$

which is possible by Theorem 1.4.4, the functions $g_k = h_k + u$ have all the required properties.

Thus we see that statements like Theorem 1.4.4 or Theorem 1.4.5 or Theorem 1.4.3′ are essentially equivalent. In the case of several variables we shall in this book prefer to work with results similar to Theorem 1.4.4, postponing the proof of results like Theorem 1.4.5 to the last chapter where we shall give an appropriate terminology for their study (cohomology groups with values in a sheaf).

1.5. The Weierstrass theorem. We give the classical formulation and ask the reader to give a statement similar to that of Theorem 1.4.3′.

Theorem 1.5.1. (*Weierstrass*) *Let z_j, $j = 1, 2, \cdots$ be a discrete sequence of different points in the open set $\Omega \subset \mathbf{C}$, and let n_j be arbitrary integers. Then there is a meromorphic function f in Ω such that f is analytic and $\neq 0$ except at the points z_j, and $f(z)(z - z_j)^{-n_j}$ is analytic and $\neq 0$ in a neighborhood of z_j for every j.*

Thus f has prescribed zeros and poles with given orders.

Proof. Let K_j be an increasing sequence of compact subsets of Ω as in the proof of Theorem 1.4.3. We shall successively choose rational functions f_j which have the desired poles and zeros in K_j, and functions $g_j \in A(\Omega)$ such that

$$(1.5.1) \qquad |f_{j+1} f_j^{-1} \exp g_j - 1| < \varepsilon_j \quad \text{on } K_j, \qquad \sum \varepsilon_j < \infty.$$

Assume that $f_1, \cdots, f_j$ and $g_1, \cdots, g_{j-1}$ have already been chosen. Let f be a rational function with the prescribed zeros and poles in K_{j+1}. Then we can write

$$f/f_j = c \prod (z - w_\nu)^{m_\nu}$$

where the product is finite, $w_\nu \in \complement K_j$ for every j, and m_ν are integers. Since no component of $\Omega \setminus K_j$ is relatively compact in Ω, we can for every ν choose $\zeta_\nu \in \complement K_{j+1}$ in the same component of $\complement K_j$ as w_ν. The function

$$f_{j+1} = f \prod (z - w_\nu')^{-m_\nu}$$

then has the appropriate zeros and poles on K_{j+1}, and

$$\log(f_{j+1}(z)/f_j(z)) = \log c + \sum m_\nu \log(z - w_\nu/z - w_\nu')$$

can be uniquely defined as an analytic function in a neighborhood of K_j since w_ν and w_ν' are in the same component of $\complement K_j$. Hence we can choose $g_j \in A(\Omega)$ so that

$$|\log(f_{j+1}/f_j) + g_j| < \log(1 + \varepsilon_j) \quad \text{in } K_j,$$

which implies (1.5.1).

From (1.5.1) it follows immediately that

$$\lim_{J \to \infty} f_{J+1} \prod_1^J e^{g_j} = f_1 \prod_1^\infty (f_{j+1} f_j^{-1} \exp g_j)$$

defines a meromorphic function f in Ω with the desired properties. In fact, the product from j to ∞ converges to an analytic function $\neq 0$ in the interior of K_j. This proves the theorem.

Corollary 1.5.2. *Every meromorphic function in Ω can be written in the form f/g where f and g are analytic in Ω.*

Proof. If the meromorphic function F has the poles z_j with orders n_j, we can use Theorem 1.5.1 to form an analytic function g with a zero of order n_j at z_j for every j. Then $Fg = f$ is analytic and $F = f/g$.

Corollary 1.5.3. *There exists a function $f \in A(\Omega)$ which cannot be continued analytically to any larger set, not even as a meromorphic function.*

More precisely: if D is a disc with center in Ω and g is a meromorphic function in D which is equal to f near the center of D, then $D \subset \Omega$.

Proof. Label all points with rational coordinates in Ω as a sequence $z_1, z_2, \cdots$ so that all such points occur an infinite number of times, and let $r_j = d(z_j, \complement\Omega)$. Choose an increasing sequence of compact sets $K_j \subset \Omega$ so that all compact subsets of Ω are contained in some K_j, and choose for each j a point $w_j \in \complement K_j$ so that $|w_j - z_j| < r_j$. Since the sequence w_j is then discrete in Ω, it follows from Theorem 1.5.1 that we can find $f \in A(\Omega)$ with a zero at w_j for every j but no other zeros. If $a \in \Omega$ has rational coordinates and $r = d(a, \complement\Omega)$, the disc $D = \{z; |z - a| < r\}$ contains infinitely many points w_j since $z_j = a$ for infinitely many j. Hence f cannot be continued to a meromorphic function in an open disc containing $\bar{D}$, for the zeros of a meromorphic function are isolated, if it is not identically 0. This proves the statement.

Corollary 1.5.4. *Let the sequence* $z_1, z_2, \cdots$ *be discrete in* Ω, *let* f_j *be analytic in a neighborhood of* z_j, *and let* n_j *be non-negative integers. Then there is a function* $f \in A(\Omega)$ *such that for every* j

$$f(z) - f_j(z) = O(|z - z_j|^{n_j + 1}), \qquad z \to z_j.$$

Proof. Let $g \in A(\Omega)$ have zeros of orders $n_j + 1$ at the points z_j. The condition on f means that $f/g - f_j/g$ shall be analytic at z_j. In view of Theorem 1.4.3 there exists a meromorphic function h with poles only at the points z_j such that $h - f_j/g$ is analytic at z_j for every j. Then $f = gh$ has the required properties.

1.6. Subharmonic functions. We recall that a C^2 function h in an open set $\Omega \subset \mathbf{C}$ is called harmonic if $\Delta h = 4\partial^2 h/\partial z \partial \bar{z} = 0$ in Ω.

Definition 1.6.1. *A function* u *defined in an open set* $\Omega \subset \mathbf{C}$ *and with values in* $[-\infty, +\infty)$ *is called subharmonic if*

(a) u *is upper semicontinuous, that is,* $\{z; z \in \Omega, u(z) < s\}$ *is open for every real number* s.

(b) *For every compact set* $K \subset \Omega$ *and every continuous function* h *on* K *which is harmonic in the interior of* K *and is* $\geq u$ *on the boundary of* K *we have* $u \leq h$ *in* K.

By our definition the function which is $-\infty$ identically is subharmonic; sometimes this is excluded in the definition.

Theorem 1.6.2. *If* u *is subharmonic and* $0 < c \in \mathbf{R}$, *it follows that* cu *is subharmonic. If* u_α, $\alpha \in A$, *is a family of subharmonic functions, then* $u = \sup_\alpha u_\alpha$ *is subharmonic if* $u < \infty$ *and* u *is upper semicontinuous, which is always the case if* A *is finite. If* $u_1, u_2, \cdots$ *is a decreasing sequence of subharmonic functions, then* $u = \lim_{j \to \infty} u_j$ *is also subharmonic.*

Proof. The first two statements are trivial. To prove the last we note that $\{z; z \in \Omega, u(z) < s\} = \cup_j \{z; z \in \Omega, u_j(z) < s\}$ which is an open set. Hence u is upper semicontinuous. If h is a majorant of u as in condition (b) and $\varepsilon > 0$, the set $\{z; z \in \partial K, u_j(z) \geq h(z) + \varepsilon\}$ is compact and decreasing. Since the limit when $j \to \infty$ is empty, the set must be empty for large j, which implies that $u_j \leq h + \varepsilon$ in K if j is large. Hence $u \leq h$ and u is subharmonic.

Other equivalent definitions of subharmonic functions are often useful:

Theorem 1.6.3. *Let* u *be defined in* Ω *with values in* $[-\infty, +\infty)$ *and assume that* u *is upper semicontinuous. Then each of the following conditions is necessary and sufficient for* u *to be subharmonic:*

(i) *If D is a closed disc $\subset \Omega$ and f is an analytic polynomial such that $u \le \operatorname{Re} f$ on ∂D, it follows that $u \le \operatorname{Re} f$ in D.*

(ii) *If $\Omega_\delta = \{z; d(z, \complement\Omega) > \delta\}$, we have*

$$u(z)2\pi \int d\mu(r) \le \int_0^{2\pi} \int u(z + re^{i\theta})\, d\theta\, d\mu(r), \qquad z \in \Omega_\delta, \tag{1.6.1}$$

for every positive measure $d\mu$ on the interval $[0,\delta]$.

(iii) *For every $\delta > 0$ and every $z \in \Omega_\delta$ there exists some positive measure $d\mu$ with support in $[0,\delta]$ such that $d\mu$ has some mass outside the origin and (1.6.1) is valid.*

Note that the integrals are well defined since u is semicontinuous.

Proof. Definition 1.6.1 implies (i), and it is also trivial that (ii) implies (iii). Thus we only have to prove that (i) $\Rightarrow$ (ii) and that (iii) implies that u is subharmonic.

(i) $\Rightarrow$ (ii). Let $z \in \Omega_\delta$ and $0 < r \le \delta$. Set $D = \{\zeta; |\zeta - z| \le r\} \subset \Omega$. If $\varphi(\theta) = \Sigma\, a_k e^{ik\theta}$ is a trigonometrical polynomial such that

$$u(z + re^{i\theta}) \le \varphi(\theta)$$

for all θ, the polynomial $f(\zeta) = a_0 + 2\,\Sigma_{k>0}\, a_k(\zeta - z)^k/r^k$ has a real part which is an upper bound for u on ∂D. Hence $u \le \operatorname{Re} f$ in D, and in particular

$$u(z) \le a_0 = (2\pi)^{-1} \int_0^{2\pi} \varphi(\theta)\, d\theta. \tag{1.6.2}$$

Now if φ is an arbitrary continuous function such that $u(z + re^{i\theta}) \le \varphi(\theta)$, we can for every $\varepsilon > 0$ find a trigonometrical polynomial φ_1 with $\varphi \le \varphi_1 \le \varphi + \varepsilon$ and conclude that (1.6.2) is valid with φ replaced by $\varphi + \varepsilon$. Hence (1.6.2) holds for every continuous function φ which is an upper bound for $u(z + re^{i\theta})$, and by definition of the integral of a semicontinuous function this proves that

$$u(z) \le (2\pi)^{-1} \int_0^{2\pi} u(z + re^{i\theta})\, d\theta.$$

Integration with respect to $d\mu(r)$ now gives (1.6.1).

(iii) implies that u is subharmonic. Let K be a compact subset of Ω and h a continuous function on K which is harmonic in the interior of K, and assume that $h \ge u$ on ∂K. If the supremum M of $v = u - h$ over K is positive, the semicontinuity of v shows that $v = M$ on a non-empty compact subset F of the interior of K. Let z_0 be a point in F with minimal distance to ∂K. If the distance is $> \delta$, then every circle $|z - z_0| = r$,

$r \le \delta$, contains points where $v(z) < M$ and, in fact, a whole arc, since v is semicontinuous. This implies that

$$\iint v(z_0 + re^{i\theta})\, d\theta\, d\mu(r) < M2\pi \int d\mu(r) = v(z_0)2\pi \int d\mu(r)$$

if $d\mu$ is a measure with the properties listed in (iii). But this contradicts the hypothesis (iii) and the fact that (1.6.1) is valid with equality for harmonic functions (because every harmonic function is subharmonic). The proof is complete.

Corollary 1.6.4. *$u_1 + u_2$ is subharmonic if u_1 and u_2 are subharmonic.*

Corollary 1.6.5. *A function u defined in an open set $\Omega \subset \mathbf{C}$ is subharmonic if every point in Ω has a neighborhood where u is subharmonic.*

In other words, subharmonicity is a local property.

Corollary 1.6.6. *If $f \in A(\Omega)$ it follows that $\log|f|$ is subharmonic in Ω.*

Proof. This follows from the maximum principle and (i) in Theorem 1.6.3.

Corollary 1.6.6 implies that $|f|$ is subharmonic if $f \in A(\Omega)$. In fact, we have

Theorem 1.6.7. *Let φ be a convex increasing function on $\mathbf{R}$ and set $\varphi(-\infty) = \lim_{x \to -\infty} \varphi(x)$. Then $\varphi(u)$ is subharmonic if u is subharmonic.*

Proof. For every x_0 there is a real number k such that

$$\varphi(x) \ge \varphi(x_0) + k(x - x_0).$$

This gives

$$(2\pi)^{-1} \int_0^{2\pi} \varphi(u(z + re^{i\theta}))\, d\theta \ge \varphi(x_0) + k((2\pi)^{-1} \int_0^{2\pi} u(z + re^{i\theta})\, d\theta - x_0).$$

If we choose x_0 so that the last parenthesis vanishes, and use the subharmonicity of u and the fact that φ is increasing, it follows that

$$\varphi(u(z)) \le \varphi((2\pi)^{-1} \int_0^{2\pi} u(z + re^{i\theta})\, d\theta) \le (2\pi)^{-1} \int_0^{2\pi} \varphi(u(z + re^{i\theta}))\, d\theta.$$

This proves the theorem, for $\varphi(u)$ is obviously semicontinuous.

Corollary 1.6.8. *Let $u_1, u_2 \ge 0$ and assume that $\log u_j$ is subharmonic in Ω, $j = 1, 2$ ($\log 0 = -\infty$). Then $\log(u_1 + u_2)$ is subharmonic in Ω.*

Proof. Let f be a polynomial in z and D a disc in Ω such that $\log(u_1 + u_2) \le \operatorname{Re} f$ on ∂D, that is, $u_1 + u_2 \le |e^f|$ on ∂D. Since

$\log u_j - \operatorname{Re} f$ is subharmonic, Theorem 1.6.7 shows that $u_j|e^{-f}|$ is subharmonic. Hence $(u_1 + u_2)|e^{-f}|$ is subharmonic and therefore ≤ 1 in D, that is, $\log(u_1 + u_2) \leq \operatorname{Re} f$ in D. This proves the corollary.

Theorem 1.6.9. *Let u be subharmonic in the open set Ω and not $-\infty$ identically in any component of Ω. Then u is integrable on all compact subsets of Ω (we write $u \in L^1_{\mathrm{loc}}(\Omega)$), which implies that $u > -\infty$ almost everywhere.*

Proof. If $z \in \Omega$, $u(z) > -\infty$, and D is a closed disc with center at z contained in Ω, we obtain from (1.6.1) and the fact that u is bounded from above in D that u is integrable over D. If E is the set of all z such that u is integrable over a neighborhood of z, it follows that $u = -\infty$ in a neighborhood of every point in $\Omega \setminus E$. Hence both E and $\Omega \setminus E$ are open so that $\Omega \setminus E$ is a union of components of Ω, all of which must be empty by hypothesis since $u = -\infty$ in $\Omega \setminus E$.

We can now give another description of subharmonic functions.

Theorem 1.6.10. *If u is subharmonic in Ω and not $-\infty$ identically in any component of Ω, we have*

$$\int u \Delta v \, d\lambda \geq 0 \quad \textit{if } v \in C_0^{\,2}(\Omega) \textit{ and } v \geq 0. \tag{1.6.3}$$

Here λ denotes the Lebesgue measure.

Proof. If $0 < r < d(\operatorname{supp} v, \complement \Omega)$, we have for every $z \in \operatorname{supp} v$

$$2\pi u(z) \leq \int_0^{2\pi} u(z + re^{i\theta}) \, d\theta.$$

Multiplication by v and integration with respect to $d\lambda$ gives

$$\int u(z)\Big(\int_0^{2\pi} v(z - re^{i\theta}) \, d\theta - 2\pi v(z)\Big) \, d\lambda(z) \geq 0.$$

If we divide by $\pi r^2/2$ and let $r \to 0$, a Taylor expansion of v shows that the inequality tends to (1.6.3).

Theorem 1.6.11. *Let $u \in L^1_{\mathrm{loc}}(\Omega)$ and assume that* (1.6.3) *holds. Then there is one and only one subharmonic function U in Ω which is equal to u almost everywhere. If φ is an integrable non-negative function of $|z|$ with compact support, we have for every $z \in \Omega$*

$$U(z) = \lim_{\delta \to 0} \int u(z - \delta z')\varphi(z') \, d\lambda(z') \Big/ \int \varphi(z') \, d\lambda(z'). \tag{1.6.4}$$

Proof. If U is subharmonic, we have by (1.6.1) for small δ

$$U(z) \leq \int U(z - \delta z')\varphi(z')\, d\lambda(z')/\int \varphi(z')\, d\lambda(z'),$$

and since U is semicontinuous from above, the upper limit of the right-hand side when $\delta \to 0$ is $\leq U(z)$. Hence (1.6.4) must hold if $u = U$ almost everywhere.

To prove the theorem we first assume that $u \in C^2(\Omega)$. Then (1.6.3) can be integrated by parts and is therefore equivalent to $\Delta u \geq 0$. Hence

$$\int (\partial^2/\partial r^2 + r^{-1}\,\partial/\partial r + r^{-2}\,\partial^2/\partial\theta^2)\, u(z + re^{i\theta})\, d\theta \geq 0, \qquad z \in \Omega_r.$$

If we write $M(r) = (2\pi)^{-1}\int_0^{2\pi} u(z + re^{i\theta})\, d\theta$, it follows that $M''(r) + r^{-1}M'(r) \geq 0$, that is, $rM'(r)$ is increasing. Since $rM'(r) \to 0$ when $r \to 0$, we get $M'(r) \geq 0$. Hence $M(0) \leq M(r)$ which proves that u is subharmonic.

Now choose a function $\varphi \in C_0^{\infty}(\mathbf{C})$ with support in the unit disc so that $\varphi \geq 0$ and φ depends only on $|z|$. Then

$$u_\delta(z) = \int u(z - \delta z')\varphi(z')\, d\lambda(z')/\int \varphi(z')\, d\lambda(z')$$

is in $C^{\infty}(\Omega_\delta)$ and $u_\delta \to u$ in L^1 norm on compact subsets of Ω when $\delta \to 0$. It is immediately verified that (1.6.3) holds in Ω_δ with u replaced by u_δ. Hence the first part of the proof shows that u_δ is subharmonic, which implies that

$$\int u_\delta(z - \varepsilon z')\varphi(z')\, d\lambda(z')/\int \varphi(z')\, d\lambda(z')$$

decreases when $\varepsilon \searrow 0$. If we let $\delta \to 0$, we conclude that $u_\varepsilon(z)$ decreases when $\varepsilon \searrow 0$. Hence

$$U(z) = \lim_{\varepsilon \to 0} u_\varepsilon(z)$$

exists and is subharmonic by Theorem 1.6.2. Since $u_\varepsilon \to u$ in $L^1_{\text{loc}}(\Omega)$ we conclude that $U = u$ almost everywhere, which completes the proof.

In particular, we have thus proved that a function $u \in C^2$ is subharmonic if and only if $\Delta u \geq 0$. When $\Delta u > 0$ we shall say that u is strictly subharmonic.

The following complement to Corollary 1.6.8 will be needed later.

Theorem 1.6.12. *If* $0 \leq f \in C^2$ *and* $\log f$ *is subharmonic, the function* $\log(1 + f)$ *is strictly subharmonic except where* $\operatorname{grad} f = \Delta f = 0$.

Proof. That $\log f$ is subharmonic means that $f\Delta f - |\operatorname{grad} f|^2 \geq 0$. If $f_1 = 1 + f$, we obtain $f_1 \Delta f_1 - |\operatorname{grad} f_1|^2 = \Delta f + (f\Delta f - |\operatorname{grad} f|^2)$ which can vanish only if $\Delta f = 0$, which implies $\operatorname{grad} f = 0$.

Finally, we shall prove a result due to Hartogs on sequences of subharmonic functions.

Theorem 1.6.13. *Let v_k be a sequence of subharmonic functions in Ω which are uniformly bounded from above on every compact subset of Ω, and assume that $\overline{\lim}_{k\to\infty} v_k(z) \leq C$ for every $z \in \Omega$. For every $\varepsilon > 0$ and every compact set $K \subset \Omega$, one can then find k_0 so that*

$$v_k(z) \leq C + \varepsilon, \qquad z \in K,\ k > k_0.$$

Proof. Since Ω can be replaced by an arbitrary relatively compact subdomain containing K, we may assume that the sequence is uniformly bounded in Ω, or even that $v_k \leq 0$ in Ω for every k. Choose $r > 0$ so that $K \subset \Omega_{3r}$. By (1.6.1) we have for every $z \in K$

$$\pi r^2 v_k(z) \leq \int_{|z-z'|<r} v_k(z')\, d\lambda(z').$$

In view of Fatou's lemma, the upper limit of the right-hand side is $\leq \pi C r^2$ when $k \to \infty$. Hence we can for every $z \in K$ choose k_0 so that

$$\int_{|z-z'|<r} v_k(z')\, d\lambda(z') \leq \pi(C + \varepsilon/2) r^2, \qquad k > k_0.$$

Since $v_k \leq 0$, we obtain, if $|z - w| < \delta < r$,

$$\pi(r+\delta)^2 v_k(w) \leq \int_{|z'-w|<r+\delta} v_k(z')\, d\lambda(z') \leq \int_{|z'-z|\leq r} v_k(z')\, d\lambda(z').$$

If δ is sufficiently small, it follows that

$$v_k(w) < C + \varepsilon \quad \text{if } k > k_0 \qquad \text{and} \quad |w - z| < \delta,$$

and since K is compact the theorem now follows from the Borel-Lebesgue lemma.

Notes. The topics discussed in this chapter are so well known that we shall give very few references. However, we wish to point out that much stronger results than Corollary 1.3.2 are known. In fact, Mergelyan [1] has proved that, when K satisfies the hypotheses of Corollary 1.3.2, then the uniform closure of the set of restrictions to K of polynomials consists of all continuous functions on K which are analytic in the interior of K. See also Wermer [1] for a simple proof using functional analysis. The analogous statements are false in the case of several variables (see Kallin [1]). Most of section 1.6 can be found in Radó [1]. An exception is Theorem 1.6.13 which is essentially due to Hartogs.

Chapter II

ELEMENTARY PROPERTIES OF FUNCTIONS OF SEVERAL COMPLEX VARIABLES

Summary. In section 2.1 we define analytic functions as the solutions of the Cauchy-Riemann equations $\bar{\partial} u = 0$. After decomposing differential forms into a sum of forms of type (p,q), we define the $\bar{\partial}$ operator for arbitrary forms. This gives a complex of differential operators, the exactness of which is proved for special cases in sections 2.3 and 2.7; the general case is studied with quite different techniques in Chapters IV and V. In sections 2.2 and 2.4 we give mostly classical results which can be proved by a simple extension of Cauchy's integral formula to product domains in $\mathbf{C}^n$. In particular, we encounter examples of domains where all analytic functions can be continued analytically to larger domains. Domains for which this is not possible are called domains of holomorphy. These are introduced in section 2.5 where we show that they are characterized by holomorph-convexity. A more explicit convexity property, so called pseudoconvexity, is introduced in section 2.6. We prove that holomorph-convexity implies pseudoconvexity, but the converse (Levi's problem) is left for Chapter IV. The last section investigates approximation of analytic functions by polynomials and in that context the exactness of the $\bar{\partial}$ sequence. It is a preparation for Chapter III, but a reader who wants to proceed directly to Chapter IV can bypass section 2.7 and the end of section 2.3 which prepares for it.

2.1 Preliminaries. Let u be a complex valued function in $C^1(\Omega)$, where Ω is an open set in $\mathbf{C}^n$, which we identify with $\mathbf{R}^{2n}$. We shall denote the real coordinates by x_j, $1 \leq j \leq 2n$, and the complex coordinates by $z_j = x_{2j-1} + ix_{2j}$, $j = 1, \cdots, n$. As in section 1.1, we can express du as a linear combination of the differentials dz_j and $d\bar{z}_j$,

$$du = \sum_1^n \partial u/\partial z_j \, dz_j + \sum_1^n \partial u/\partial \bar{z}_j \, d\bar{z}_j, \tag{2.1.1}$$

where we have used the notation

$$\partial u/\partial z_j = \tfrac{1}{2}(\partial u/\partial x_{2j-1} - i\partial u/\partial x_{2j}), \tag{2.1.2}$$
$$\partial u/\partial \bar{z}_j = \tfrac{1}{2}(\partial u/\partial x_{2j-1} + i\partial u/\partial x_{2j}).$$

With the notation

$$\partial u = \sum_1^n \partial u/\partial z_j \, dz_j, \qquad \bar{\partial} u = \sum_1^n \partial u/\partial \bar{z}_j \, d\bar{z}_j, \tag{2.1.3}$$

we may also write (2.1.1) in the form

$$du = \partial u + \bar{\partial} u. \tag{2.1.1$'$}$$

Differential forms which are linear combinations of the differentials dz_j are said to be of type (1,0), and those which are linear combinations of $d\bar{z}_j$ are said to be of type (0,1). Thus ∂u (resp. $\bar{\partial} u$) is the component of du of type (1,0) (resp. (0,1)).

Definition 2.1.1. *A function $u \in C^1(\Omega)$ is said to be analytic (or holomorphic) in Ω if du is of type (1,0), that is, if $\bar{\partial} u = 0$ (the Cauchy–Riemann equations). The set of all analytic functions in Ω is denoted by $A(\Omega)$.*

The differential operators ∂ and $\bar{\partial}$ are obviously linear and satisfy the product rule. Hence $A(\Omega)$ is a ring.

Now let u be an analytic function in Ω with values in $\mathbf{C}^\nu$, that is,

$$u = (u_1, \cdots, u_\nu),$$

where each component u_j is analytic in Ω. If $v \in C^1(\omega)$ for some open set ω containing the range of u, the function

$$u^*v : \Omega \ni z \to v(u(z))$$

is in $C^1(\Omega)$ and we have

$$d(u^*v) = \sum_1^\nu \partial v/\partial u_j \, du_j + \sum_1^\nu \partial v/\partial \bar{u}_j \, d\bar{u}_j.$$

Since du_j is of type (1,0) and $d\bar{u}_j$ of type (0,1) in Ω, it follows that

$$\partial(u^*v) = \sum_1^\nu \partial v/\partial u_j \, du_j, \qquad \bar{\partial}(u^*v) = \sum_1^\nu \partial v/\partial \bar{u}_j \, d\bar{u}_j.$$

Hence u^*v is analytic if v is analytic. More generally, the decomposition of d as $\partial + \bar{\partial}$ and the notion of analytic function are invariant under analytic maps.

The implicit function theorem extends immediately to analytic functions:

Theorem 2.1.2. *Let* $f_j(w,z)$, $j = 1, \cdots, m$, *be analytic functions of* $(w,z) = (w_1, \cdots, w_m, z_1, \cdots, z_n)$ *in a neighborhood of a point* (w^0,z^0) *in* $\mathbf{C}^m \times \mathbf{C}^n$, *and assume that* $f_j(w^0,z^0) = 0$, $j = 1, \cdots, m$ *and that*

$$\det(\partial f_j/\partial w_k)_{j,k=1}^m \neq 0 \qquad \text{at } (w^0,z^0).$$

Then the equations $f_j(w,z) = 0$, $j = 1, \cdots, m$, *have a uniquely determined analytic solution* $w(z)$ *in a neighborhood of* z^0, *such that* $w(z^0) = w^0$.

Proof. We may regard the equations $f_j(w, z) = 0$ as $2m$ real equations $\operatorname{Re} f_j(w,z) = 0$ and $\operatorname{Im} f_j(w, z) = 0$ for the $2m$ real unknowns $\operatorname{Re} w_k$ and $\operatorname{Im} w_k$. In order to apply the usual implicit function theorem, we have to prove that at (w^0,z^0) the equations $df_j = 0$ and $dz_k = 0$, $j = 1, \cdots, m$, $k = 1, \cdots, n$, imply $dw_j = 0, j = 1, \cdots, m$. But this is obvious since the determinant of the system

$$\sum_1^m \partial f_j/\partial w_k \, dw_k = 0, \qquad j = 1, \cdots, m$$

is not 0. Hence the equations $f_j(w,z) = 0$ define uniquely C^1 functions w_k in a neighborhood of z^0 such that $w(z^0) = w^0$. To prove the analyticity we only note that for these functions w_k

$$\sum_1^m \partial f_j/\partial w_k \, dw_k + \sum_1^n \partial f_j/\partial z_k \, dz_k = df_j = 0$$

and we can solve this system of equations for dw_k and find that dw_k is a linear combination of $dz_1, \cdots, dz_n$.

Note that Theorem 2.1.2 means in particular that an analytic map of $\mathbf{C}^n$ into itself has locally an analytic inverse where the Jacobian does not vanish.

Finally we shall extend the definition of the ∂ and $\bar\partial$ operators to arbitrary differential forms. A differential form f is said to be of type (p, q) if it can be written in the form

$$f = \sum_{|I|=p} \sum_{|J|=q} f_{I,J} \, dz^I \wedge d\bar z^J,$$

where $I = (i_1, \cdots, i_p)$ and $J = (j_1, \cdots, j_q)$ are multi-indices, that is, sequences of indices between 1 and n. We have here used the notation

$$dz^I \wedge d\bar z^J = dz_{i_1} \wedge \cdots \wedge dz_{i_p} \wedge d\bar z_{j_1} \wedge \cdots \wedge d\bar z_{j_q}.$$

Every differential form can be written in one and only one way as a sum of forms of type (p,q); $0 \leq p, q \leq n$. If f is of type (p,q), the exterior differential

$$df = \sum df_{I,J} \wedge dz^I \wedge d\bar{z}^J$$

can be written $df = \partial f + \bar{\partial} f$, where

$$\partial f = \sum_{I,J} \partial f_{I,J} \wedge dz^I \wedge d\bar{z}^J, \qquad \bar{\partial} f = \sum_{I,J} \bar{\partial} f_{I,J} \wedge dz^I \wedge d\bar{z}^J$$

are of type $(p + 1,q)$ and $(p,q + 1)$, respectively. Since $0 = d^2 f = \partial^2 f + (\partial\bar{\partial} + \bar{\partial}\partial) f + \bar{\partial}^2 f$ and all terms are of different types, we obtain

$$\partial^2 = 0, \qquad \partial\bar{\partial} + \bar{\partial}\partial = 0, \qquad \bar{\partial}^2 = 0. \tag{2.1.4}$$

Hence the equation

$$\bar{\partial} u = f, \tag{2.1.5}$$

where f is of type $(p,q + 1)$, cannot have a solution u unless

$$\bar{\partial} f = 0.$$

This shows that even if we are primarily interested in the Cauchy–Riemann equations (2.1.5) for functions u only, it is natural to study the $\bar{\partial}$ operator also for forms of type (0,1), and therefore for forms of type (0,2), $\cdots$.

If u is a holomorphic map of $\Omega \subset \mathbf{C}^n$ into $\mathbf{C}^\nu$ and if $f = \sum f_{I,J}\, du^I \wedge d\bar{u}^J$ is a form defined in an open neighborhood of the range of u, we can define a form $u^* f$ in Ω by

$$u^* f = \sum f_{I,J}(u(z))\, du^I \wedge d\bar{u}^J,$$

where du_k and $d\bar{u}_k$ for $k = 1, \cdots, \nu$ are differential forms in Ω of type (1,0) and (0,1), respectively, since u_k is analytic. Hence $u^* f$ is of type (p,q) if f is of type (p,q), and since $d(u^* f) = u^*(df)$ it follows that

$$\partial(u^* f) = u^*(\partial f), \qquad \bar{\partial}(u^* f) = u^*(\bar{\partial} f).$$

If $\mathscr{F}$ is a space of functions (or distributions) we shall use the notation $\mathscr{F}_{(p,q)}$ for the space of forms of type (p,q) with coefficients belonging to $\mathscr{F}$.

2.2. Applications of Cauchy's integral formula in polydiscs. A set $D \subset \mathbf{C}^n$ is called a polydisc if there are discs $D_1, \cdots, D_n$ in $\mathbf{C}$ such that

$$D = \prod_1^n D_j = \{z; z_j \in D_j, j = 1, \cdots, n\}.$$

The set $\Pi_1^n \partial D_j$ is called the distinguished boundary of D and we denote it by $\partial_0 D$.

Theorem 2.2.1. *Let D be an open polydisc and let u be a continuous function in $\bar{D}$ which in D is an analytic function of each z_j when the other variables are kept fixed. Then we have*

(2.2.1) $u(z) =$

$$(2\pi i)^{-n} \int_{\partial_0 D} u(\zeta_1, \cdots, \zeta_n)(\zeta_1 - z_1)^{-1} \cdots (\zeta_n - z_n)^{-1} \, d\zeta_1 \cdots d\zeta_n .$$

Hence $u \in C^\infty(D)$ and u is in fact analytic in D.

Proof. From Corollary 1.2.5 and the continuity of u, it follows that u is an analytic function of $z_j \in D_j$ if the other coordinates z_k are given arbitrary fixed values in $\bar{D}_k$. By repeated use of (1.2.3), we therefore obtained (2.2.1), and, since the integrand is a C^∞ analytic function of z when $(\zeta, z) \in \partial_0 D \times D$, the theorem follows.

Corollary 2.2.2. *If Ω is an open set in $\mathbf{C}^n$ and $u \in A(\Omega)$, it follows that $u \in C^\infty(\Omega)$ and that all derivatives of u are also analytic in Ω.*

We can also immediately obtain bounds for the derivatives of u. In doing so we shall call an n-tuple $\alpha = (\alpha_1, \cdots, \alpha_n)$ of non-negative integers a multi-order and write $\partial^\alpha = (\partial/\partial z_1)^{\alpha_1} \cdots (\partial/\partial z_n)^{\alpha_n}$. The operator $\bar{\partial}^\alpha$ is defined similarly and we write $\alpha! = \alpha_1! \cdots \alpha_n!$ and $|\alpha| = \alpha_1 + \cdots + \alpha_n$.

Theorem 2.2.3. *For every compact set $K \subset \Omega$ (open set in $\mathbf{C}^n$) and every open neighborhood ω of K there are constants C_α for all multi-orders α such that*

$$\sup_K |\partial^\alpha u| \leq C_\alpha \|u\|_{L^1(\omega)}, \qquad u \in A(\Omega). \tag{2.2.2}$$

Proof. This is proved by repeated use of Theorem 1.2.4 if ω is a polydisc and K is a compact subset of ω. Since K can be covered by a finite number of compact subsets of polydiscs contained in ω, the theorem follows.

Corollary 2.2.4. *If $u_k \in A(\Omega)$ and $u_k \to u$ when $k \to \infty$, uniformly on compact subsets of Ω, it follows that $u \in A(\Omega)$.*

Proof. Repetition of that of Corollary 1.2.5.

Corollary 2.2.5. *If $u_k \in A(\Omega)$ and the sequence $|u_k|$ is uniformly bounded on every compact subset of Ω, there is a subsequence u_{k_j} converging uniformly on every compact subset of Ω to a limit $u \in A(\Omega)$.*

Proof. Repetition of that of Corollary 1.2.6.

We shall now consider power series expansions of functions which are analytic in polydiscs. In doing so we say that a series $\Sigma_\alpha a_\alpha(z)$ converges *normally* in an open set Ω if $\Sigma_\alpha \sup_K |a_\alpha(z)|$ converges for every compact set $K \subset \Omega$. This implies of course that $\Sigma_\alpha a_\alpha(z)$ exists and is independent of the order of summation and that the sum is analytic if all a_α are analytic.

Theorem 2.2.6. *If* u *is analytic in the polydisc* $D = \{z; |z_j| < r_j, j = 1, \cdots, n\}$, *we have*

$$u(z) = \sum_\alpha \partial^\alpha u(0)/\alpha! z^\alpha, \qquad z \in D,$$

with normal convergence.

Proof. The power series expansion

$$(\zeta_1 - z_1)^{-1} \cdots (\zeta_n - z_n)^{-1} = \sum z^\alpha / \zeta^\alpha \zeta_1 \cdots \zeta_n$$

converges normally when $(\zeta, z) \in \partial_0 D \times D$. Hence we can multiply by $u(\zeta_1, \cdots, \zeta_n)$ and integrate term by term in (2.2.1) if u is continuous in $\bar{D}$. Since differentiation of (2.2.1) gives

$$\partial^\alpha u(0) = (2\pi i)^{-n} \alpha! \int_{\partial_0 D} u(\zeta_1, \cdots, \zeta_n) \prod_1^n \zeta_j^{-\alpha_j - 1} \, d\zeta_1 \cdots d\zeta_n, \tag{2.2.3}$$

the theorem follows if u is continuous in $\bar{D}$. In general, we only have to apply this result to polydiscs which are relatively compact in D.

We also note that the proof gives

Theorem 2.2.7 (Cauchy's inequalities). *If* u *is analytic and* $|u| \leq M$ *in the polydisc* $\{z; |z_j| < r_j, j = 1, \cdots, n\}$, *it follows that*

$$|\partial^\alpha u(0)| \leq M \alpha! r^{-\alpha}. \tag{2.2.4}$$

Proof. This follows from (2.2.3) applied to smaller polydiscs.

From Theorem 2.2.6 we obtain the uniqueness of analytic continuation as in the proof of Corollary 1.2.9. (We could also apply Corollary 1.2.9.) The maximum principle also extends immediately from one to several variables.

We shall end this section by proving the Hartogs theorem that a separately analytic function is analytic. A corresponding result would be false for functions of real variables: the function $f(x,y) = xy/(x^2 + y^2)$, $f(0,0) = 0$, is infinitely differentiable with respect to x (or y) when y (or x) is kept fixed, but in spite of that f is not even continuous at the origin.

Theorem 2.2.8. *If u is a complex valued function defined in the open set $\Omega \subset \mathbf{C}^n$ and u is analytic in each variable z_j when the other variables are given arbitrary fixed values, then u is analytic in Ω.*

The statement is purely local, so we only have to prove it in polydiscs.

Lemma 2.2.9. *If u satisfies the hypotheses of Theorem 2.2.8 in the polydisc $\Omega = \{z; |z_j| < r_j, j = 1, \cdots, n\}$ and if $|u|$ is bounded in Ω, it follows that u is analytic in Ω.*

Proof. In view of Theorem 2.2.1 it is sufficient to prove that u is continuous in Ω. Let $|u| \le M$ in Ω. Then we claim that

$$|u(z) - u(\zeta)| \le 2M \sum_1^n r_j|z_j - \zeta_j|/|r_j^2 - z_j\bar{\zeta}_j| \quad \text{if } z, \zeta \in \Omega. \tag{2.2.5}$$

Since

$$u(z) - u(\zeta) = \sum_1^n (u(\zeta_1, \cdots, \zeta_{j-1}, z_j, \cdots, z_n) - u(\zeta_1, \cdots, \zeta_j, z_{j+1}, \cdots, z_n)),$$

it is sufficient to prove the estimate in the case of one variable. But then it follows from the Schwarz lemma, since the function $z \to u(z) - u(\zeta)$ is $\le 2M$ in absolute value and vanishes when $z = \zeta$. Hence (2.2.5) is valid, which proves the lemma.

In what follows we assume that Theorem 2.2.8 has already been proved for functions of fewer than n variables. (Note that it is trivial when $n = 1$.)

Lemma 2.2.10. *Let the hypotheses of Theorem 2.2.8 be fulfilled and let $D = \Pi_1^n D_j$ be a closed polydisc with non-empty interior contained in Ω. Then there exist discs $D_j' \subset D_j$ with positive radii and $D_n' = D_n$ such that u is bounded in $D' = \Pi_1^n D_j'$, hence analytic in the interior of D'.*

Proof. Let

$$E_M = \{z'; z' \in \prod_1^{n-1} D_j \quad \text{and} \quad |u(z',z_n)| \le M \quad \text{when } z_n \in D_n\}.$$

Then E_M is closed since, by the inductive hypothesis, u is analytic and therefore continuous in z' for fixed z_n. Further $\cup_1^\infty E_M = \Pi_1^{n-1} D_j$. Hence Baire's theorem shows that E_M has an interior point for large M, and if we choose D' so that $D' \subset E_M \times D_n$, the lemma is proved.

Lemma 2.2.11. *Let u be a complex valued function in the polydisc $D = \{z; |z_j - z_j^0| < R, j = 1, \cdots, n\}$, assume that u is analytic in*

$z' = (z_1, \cdots, z_{n-1})$ if z_n is fixed and that u is analytic and bounded in $D' = \{z; |z_j - z_j^0| < r, j = 1, \cdots, n-1, |z_n - z_n^0| < R\}$ for some $r > 0$. Then u is analytic in D.

Proof. We may assume that $z^0 = 0$. Choose R_1 and R_2 with $0 < R_1 < R_2 < R$. By Theorem 2.2.6 we have

$$u(z) = \sum_{\alpha} a_\alpha(z_n) z'^\alpha, \qquad z \in D, \tag{2.2.6}$$

where α only runs through multi-orders with $n-1$ places,

$$a_\alpha(z_n) = \partial^\alpha u(0, z_n)/\alpha!$$

is analytic in z_n and

$$|a_\alpha(z_n)| R_2^{|\alpha|} \to 0 \quad \text{when } |\alpha| \to \infty \text{ for fixed } z_n \text{ with } |z_n| < R. \tag{2.2.7}$$

Further, Cauchy's inequality gives

$$|a_\alpha(z_n)| r^{|\alpha|} \le M \tag{2.2.8}$$

if M is a bound for $|u|$ in D'. Now apply Theorem 1.6.13 to the subharmonic functions

$$z_n \to \frac{1}{|\alpha|} \log |a_\alpha(z_n)|.$$

In view of (2.2.8), these functions are uniformly bounded from above when $|z_n| < R$ and by (2.2.7) the upper limit when $|\alpha| \to \infty$ is $\le \log(1/R_2)$ for fixed z_n. Hence Theorem 1.6.13 shows that for large $|\alpha|$

$$\frac{1}{|\alpha|} \log |a_\alpha(z_n)| \le \log \frac{1}{R_1} \quad \text{if } |z_n| < R_1,$$

that is,

$$|a_\alpha(z_n)| R_1^{|\alpha|} \le 1 \quad \text{for large } |\alpha| \text{ if } |z_n| < R_1.$$

This proves that the series (2.2.6) converges normally in D, and, since the terms are analytic, we conclude that u is analytic.

Proof of Theorem 2.2.8. Given $\zeta \in \Omega$, we choose $R > 0$ so that the polydisc $\{z; |z_j - \zeta_j| \le 2R, j = 1, \cdots, n\}$ is contained in Ω. By Lemma 2.2.10 we can find z^0 with $\max_j |z_j^0 - \zeta_j| < R$ so that the hypotheses of Lemma 2.2.11 are fulfilled for some $r > 0$. Hence u is analytic in a neighborhood of ζ, which proves the theorem.

2.3. The inhomogeneous Cauchy–Riemann equations in a polydisc. We first consider the equation

$$\bar{\partial} u = f$$

where f is a given form of type (0,1) with *compact support*, and the unknown u is a function. We recall that $\bar{\partial} f = 0$ is a necessary condition for the existence of a solution. Explicitly, this means that we want to solve the overdetermined system of differential equations

$$\partial u / \partial \bar{z}_j = f_j, \qquad j = 1, \cdots, n, \tag{2.3.1}$$

when the compatibility conditions

$$\partial f_j / \partial \bar{z}_k - \partial f_k / \partial \bar{z}_j = 0, \qquad j,k = 1, \cdots, n \tag{2.3.2}$$

are fulfilled.

Theorem 2.3.1. *Let $f_j \in C_0^k(\mathbf{C}^n)$, $j = 1, \cdots, n$, where $k > 0$, and assume that* (2.3.2) *is fulfilled* ($n > 1$). *Then there is a function $u \in C_0^k(C^n)$ satisfying* (2.3.1).

Note that the theorem is false when $n = 1$ (take an arbitrary $f_1 \in C_0^\infty$ with Lebesgue integral different from 0).

Proof. We set

$$\begin{aligned} u(z) &= (2\pi i)^{-1} \iint (\tau - z_1)^{-1} f_1(\tau, z_2, \cdots, z_n)\, d\tau \wedge d\bar{\tau} \\ &= -(2\pi i)^{-1} \iint \tau^{-1} f_1(z_1 - \tau, z_2, \cdots)\, d\tau \wedge d\bar{\tau}. \end{aligned}$$

The second form of the definition shows that $u \in C^k(\mathbf{C}^n)$, and it is clear that $u(z) = 0$ if $|z_2| + \cdots + |z_n|$ is large enough. From Theorem 1.2.2 it follows that $\partial u / \partial \bar{z}_1 = f_1$. If $k > 1$, by differentiating under the sign of integration and using the fact that $\partial f_1 / \partial \bar{z}_k = \partial f_k / \partial \bar{z}_1$, we obtain

$$\partial u / \partial \bar{z}_k = (2\pi i)^{-1} \iint (\tau - z_1)^{-1}\, \partial f_k(\tau, z_2, \cdots) / \partial \bar{\tau}\, d\tau \wedge d\bar{\tau} = f_k(z),$$

where the last equality follows from Theorem 1.2.1. Hence u satisfies all the equations (2.3.1), which means in particular that u is analytic outside a compact set. From the uniqueness of analytic continuation, we now conclude that u has compact support.

Theorem 2.3.1 leads immediately to the following theorem of Hartogs.

Theorem 2.3.2. *Let Ω be an open set in $\mathbf{C}^n$, $n > 1$, and let K be a compact subset of Ω such that $\Omega \setminus K$ is connected. For every $u \in A(\Omega \setminus K)$ one can then find $U \in A(\Omega)$ so that $u = U$ in $\Omega \setminus K$.*

Proof. Let $\varphi \in C_0^\infty(\Omega)$ be equal to 1 in a neighborhood of K. Set $u_0 = (1 - \varphi)u$, defined as 0 in K. Then $u_0 \in C^\infty(\Omega)$, and we want to find $v \in C^\infty(\mathbf{C}^n)$ so that

$$U = u_0 - v$$

has the required properties. The function U will be analytic if

$$\bar{\partial} v = \bar{\partial} u_0 = -u\bar{\partial}\varphi = f, \tag{2.3.3}$$

where f, defined as 0 in K and outside Ω, has components in $C_0^\infty(\mathbf{C}^n)$. Hence the equation (2.3.3) has a solution which vanishes in the unbounded component of the complement of the support of φ. The boundary of this set belongs to $\Omega \setminus K$, so there exists an open set in $\Omega \setminus K$ where $v = 0$ and $u = u_0$. Hence the analytic function U in Ω which we have defined coincides with u on some open subset of $\Omega \setminus K$, and since this is a connected set we have $u = U$ in $\Omega \setminus K$.

Every function which is analytic in $\Omega \setminus K$ can thus be continued analytically to the larger set Ω. This is in striking contrast with the situation in the case of one complex variable (Corollary 1.5.3). We shall find another example of this phenomenon in the next section and shall make a thorough study of it in section 2.5.

Without changing the method of proof we can also give a stronger version of Theorem 2.3.2 where the function u to be extended is just defined on $\partial\Omega$.

Theorem 2.3.2′. *Let Ω be a bounded open set in $\mathbf{C}^n$, $n > 1$, such that $\complement\bar{\Omega}$ is connected and $\partial\Omega \in C^4$. Denote by ρ a real valued function in C^4 such that ρ vanishes precisely on $\partial\Omega$ and* grad $\rho \neq 0$ *on $\partial\Omega$. If $u \in C^4(\bar{\Omega})$ and $\bar{\partial} u \wedge \bar{\partial}\rho = 0$ on $\partial\Omega$, one can then find an analytic function $U \in C^1(\bar{\Omega})$ such that $U = u$ on $\partial\Omega$.*

Before the proof we make a few remarks. First note that, conversely, the existence of the analytic function U implies that $U - u = \rho h$ with $h \in C^0(\bar{\Omega})$, hence $\bar{\partial} U - \bar{\partial} u = h\bar{\partial}\rho$ on $\partial\Omega$ so that $\bar{\partial} u \wedge \bar{\partial}\rho = \bar{\partial} U \wedge \bar{\partial}\rho = 0$ on $\partial\Omega$. The hypothesis $\bar{\partial} u \wedge \bar{\partial}\rho = 0$ can also be stated as follows:

$$\sum_1^n t_j\, \partial u/\partial \bar{z}_j = 0 \quad \text{if} \sum_1^n t_j\, \partial\rho/\partial\bar{z}_j = 0,$$

that is, u shall satisfy all tangential Cauchy–Riemann equations. Note that these involve only the values of u on $\partial\Omega$. The differentiability assumptions can easily be reduced by two units.

Proof of Theorem 2.3.2′. We shall first construct $U_0 \in C^2(\bar{\Omega})$ so that $U_0 = u$ on $\partial\Omega$ and

$$\bar{\partial} U_0 = O(\rho^2) \quad \text{at } \partial\Omega.$$

To do so we note that by assumption

$$\bar{\partial} u = h_0 \, \bar{\partial}\rho + \rho h_1$$

for suitable $h_0 \in C^3(\bar{\Omega})$ and $h_1 \in C^2_{(0,1)}(\bar{\Omega})$. Hence

$$\bar{\partial}(u - h_0\rho) = \rho(h_1 - \bar{\partial} h_0) = \rho h_2, \quad \text{where } h_2 \in C^2_{(0,1)}(\bar{\Omega}).$$

Since $0 = \bar{\partial}(\rho h_2) = \bar{\partial}\rho \wedge h_2 + \rho \bar{\partial} h_2$, we have $\bar{\partial}\rho \wedge h_2 = 0$ on $\partial\Omega$. We can therefore write $h_2 = h_3 \bar{\partial}\rho + \rho h_4$ with $h_3 \in C^2(\bar{\Omega})$ and $h_4 \in C^1_{(0,1)}(\bar{\Omega})$. With $U_0 = u - h_0\rho - h_3\rho^2/2$, we then obtain

$$\bar{\partial} U_0 = \rho^2(h_4 - \bar{\partial} h_3/2),$$

which completes the construction of U_0. Now set

$$f = \bar{\partial} U_0 \text{ in } \Omega, \qquad f = 0 \text{ in } \complement\Omega.$$

The form f is then in $C^1_{(0,1)}(\mathbf{C}^n)$ and has compact support, so by Theorem 2.3.1 we can choose a function $v \in C^1(\mathbf{C}^n)$ so that $\bar{\partial} v = f$ and v has compact support. Since v is analytic in $\complement\bar{\Omega}$ and this is a connected set, it follows that $v = 0$ in $\complement\bar{\Omega}$. The function $U = U_0 - v$ is therefore equal to $U_0 = u$ on $\partial\Omega$, and $\bar{\partial} U = \bar{\partial} U_0 - \bar{\partial} v = f - f = 0$ in Ω. This completes the proof.

The study of the equation $\bar{\partial} u = f$ is slightly more complicated even in polydiscs when f does not have compact support. Since we shall need results when f is a form of type $(0,q + 1)$ for an arbitrary $q \geq 0$ in section 2.6, we allow f to be of type $(p,q + 1)$ for arbitrary $p,q \geq 0$. For simplicity we assume that the components are in C^∞.

Theorem 2.3.3. *Let D be an open polydisc and let $f \in C^\infty_{(p,q+1)}(D)$ $(p,q \geq 0)$ satisfy the condition $\bar{\partial} f = 0$. If $D' \subset\subset D$ (that is, D' is relatively compact in D), we can find $u \in C^\infty_{(p,q)}(D')$ with $\bar{\partial} u = f$ in D'.*

Proof. We shall prove inductively that the theorem is true if f does not involve $d\bar{z}_{k+1}, \cdots, d\bar{z}_n$. This is trivial if $k = 0$, for f must then be 0 since every term in f is of degree $q + 1 > 0$ with respect to $d\bar{z}$. For $k = n$, the statement is identical to the theorem. Assuming that it has already been proved when k is replaced by $k - 1$, we write

$$f = d\bar{z}_k \wedge g + h,$$

where $g \in C^{\infty}_{(p,q)}(D)$, $h \in C^{\infty}_{(p,q+1)}(D)$, and g and h are independent of $d\bar{z}_k, \cdots, d\bar{z}_n$. Write

$$g = \sum_{|I|=p}' \sum_{|J|=q}' g_{I,J}\, dz^I \wedge d\bar{z}^J$$

where Σ' means that we sum only over increasing multi-indices. Since $\bar{\partial} f = 0$, we obtain

$$\partial g_{I,J}/\partial \bar{z}_j = 0, \qquad j > k, \tag{2.3.4}$$

for this is apart from a factor ± 1 the coefficient of $dz^I \wedge d\bar{z}^J \wedge d\bar{z}_k \wedge d\bar{z}_j$ in $\bar{\partial} f$. Thus $g_{I,J}$ is analytic in these variables.

We now choose a solution $G_{I,J}$ of the equation

$$\partial G_{I,J}/\partial \bar{z}_k = g_{I,J}. \tag{2.3.5}$$

To do so, we choose $\psi \in C_0{}^{\infty}(D_k)$ so that $\psi(z_k) = 1$ in a neighborhood $D'' \subset D$ of $\bar{D}'$, and set

$$G_{I,J}(z) =$$

$$(2\pi i)^{-1} \iint (\tau - z_k)^{-1} \psi(\tau) g_{I,J}(z_1, \cdots, z_{k-1}, \tau, z_{k+1}, \cdots, z_n)\, d\tau \wedge d\bar{\tau}$$

$$= -(2\pi i)^{-1} \iint \tau^{-1} \psi(z_k - \tau) g_{I,J}(z_1, \cdots, z_{k-1}, z_k - \tau, z_{k+1}, \cdots, z_n)\, d\tau \wedge d\bar{\tau}.$$

The last expression shows that $G_{I,J} \in C^{\infty}(D)$, from Theorem 1.2.2 it follows that (2.3.5) holds in D'', and in view of (2.3.4) we obtain by differentiating under the sign of integration

$$\partial G_{I,J}/\partial \bar{z}_j = 0, \qquad j > k. \tag{2.3.6}$$

If we set

$$G = \sum_{I,J}' G_{I,J}\, dz^I \wedge d\bar{z}^J,$$

it follows that in D''

$$\bar{\partial} G = \sum_{I,J}' \sum_j \partial G_{I,J}/\partial \bar{z}_j\, d\bar{z}_j \wedge dz^I \wedge d\bar{z}^J = d\bar{z}_k \wedge g + h_1,$$

where h_1 is the sum when j runs from 1 to $k - 1$ and is independent of $d\bar{z}_k, \cdots, d\bar{z}_n$. Hence $h - h_1 = f - \bar{\partial} G$ does not involve $d\bar{z}_k, \cdots, d\bar{z}_n$, so by the inductive hypothesis we can find $v \in C^{\infty}_{(p,q)}(D')$ so that $\bar{\partial} v = f - \bar{\partial} G$ there. (Note that $\bar{\partial}(f - \bar{\partial} G) = \bar{\partial} f = 0$.) But then $u = v + G$ satisfies the equation $\bar{\partial} u = f$, which completes the proof.

By arguing as in the proof of Theorem 1.4.4, it is not difficult to prove that there is a solution $u \in C^{\infty}_{(p,q)}(D)$ of the equation $\bar{\partial} u = f$ when the hypotheses of Theorem 2.3.3 are fulfilled. However, we shall not do so until section 2.7, when we will have extended Theorem 2.3.3 to more general sets than polydiscs.

2.4. Power series and Reinhardt domains. Let a_α be complex numbers, defined for all multi-orders α, and consider the power series in $\mathbf{C}^n$

$$\sum_\alpha a_\alpha z^\alpha.$$

We define its *domain of convergence* D as the set of all z such that the series is absolutely convergent at every point in a neighborhood of z. Further we denote by B the set of all z such that $|a_\alpha z^\alpha| \le C$ for all α and some constant C. It is obvious that D belongs to the interior B° of B. To prove that there is equality we first prove Abel's lemma.

Lemma 2.4.1. *If $w \in B$, the power series converges normally in the polydisc $|z_j| < |w_j|, j = 1, \cdots, n$.*

Proof. By hypothesis there is a constant C such that $|a_\alpha w^\alpha| \le C$. If $|z_j| \le k_j |w_j|$ and $k_j < 1$ for every j, we obtain $|a_\alpha z^\alpha| \le C\, k^\alpha$, and since $\Sigma_\alpha k^\alpha = \Pi_1^n (1 - k_j)^{-1} < \infty$, the lemma is proved.

Theorem 2.4.2. *The convergence domain D is the interior of B, and the power series converges normally in D so that the sum is analytic in D.*

Proof. This is an immediate consequence of Lemma 2.4.1.

Theorem 2.4.3. *Let $D^* = \{\xi ; \xi \in \mathbf{R}^n, (e^{\xi_1}, \cdots, e^{\xi_n}) \in D\}$. Then D^* is an open convex set in $\mathbf{R}^n$, and if $\xi \in D^*$ it follows that $\eta \in D^*$ if $\eta_j \le \xi_j$ for every j. Further, we have $z \in D$ if and only if $|z_j| \le e^{\xi_j}, j = 1, \cdots, n$, for some $\xi \in D^*$.*

Proof. Define B^* by substituting B for D in the definition of D^*. From Theorem 2.4.2, it follows that D^* is the interior of B^*. Now if $\xi, \eta \in B^*$, one can find C so that

$$|a_\alpha| \exp\Big(\sum_1^n \alpha_j \xi_j\Big) \le C \quad \text{and} \quad |a_\alpha| \exp\Big(\sum_1^n \alpha_j \eta_j\Big) \le C \quad \text{for all } \alpha.$$

If $\lambda, \mu \ge 0$ and $\lambda + \mu = 1$, it follows that

$$|a_\alpha| \exp\Big(\sum_1^n \alpha_j(\lambda \xi_j + \mu \eta_j)\Big) \le C.$$

Hence $\lambda\xi + \mu\eta \in B^*$, that is, B^* is convex. This gives immediately the asserted properties of D^*.

Conversely, if $D^* \subset \mathbf{R}^n$ has the properties listed in Theorem 2.4.3 and we define D according to the last statement there, we obtain an open set D which is precisely the domain of convergence for some power series. This follows from Theorem 2.4.5 below and the fact which will be proved in section 2.5 that there exists an analytic function in D which cannot be continued analytically beyond D. (See Corollary 2.5.8 below.)

Theorem 2.4.3 means in particular that the domain of convergence of a power series has the property in the following definition:

Definition 2.4.4. *An open set $\Omega \subset \mathbf{C}^n$ is called a Reinhardt domain if $(z_1, \cdots, z_n) \in \Omega$ implies $(e^{i\theta_1}z_1, \cdots, e^{i\theta_n}z_n) \in \Omega$ for arbitrary real $\theta_1, \cdots, \theta_n$.*

Theorem 2.4.5. *Let $\Omega \subset \mathbf{C}^n$ be a connected Reinhardt domain containing the origin, and let $f \in A(\Omega)$. Then there exists one (and only one) power series such that*

$$f(z) = \sum_\alpha a_\alpha z^\alpha$$

with normal convergence in Ω.

Proof. The uniqueness is obvious, since differentiation term by term (permissible in view of Theorem 2.2.3) gives that $a_\alpha = \partial^\alpha f(0)/\alpha!$ To prove the existence we take an arbitrary number $\varepsilon > 0$ and define Ω_ε as the set of all $z \in \Omega$ such that $d(z, \complement\Omega) > \varepsilon|z|$. ($|z|$ is some norm in $\mathbf{C}^n$.) We have $0 \in \Omega_\varepsilon$, and Ω_ε is open. Let Ω_ε' be the component of Ω_ε which contains the origin. It is clear that Ω_ε' increases when $\varepsilon \searrow 0$, and since every $z \in \Omega$ can be joined to 0 with a polygon in Ω, we have $\Omega = \cup_{\varepsilon > 0}\Omega_\varepsilon'$. When $z \in \Omega_\varepsilon'$, we set

$$g(z) = (2\pi i)^{-n} \int_{\partial_0 T_\varepsilon} f(t_1z_1, \cdots, t_nz_n)(t_1 - 1)^{-1} \cdots (t_n - 1)^{-1}\, dt_1 \cdots dt_n,$$

where $T_\varepsilon = \{t; |t_j| \le 1 + \varepsilon, j = 1, \cdots, n\}$. The integral is defined, for the distance from z to $(1 + \varepsilon)z$ is $\varepsilon|z|$, so that, if $z \in \Omega_\varepsilon'$, it follows that $(1 + \varepsilon)z \in \Omega$. Since Ω is a Reinhardt domain, this implies that $(t_1z_1, \cdots, t_nz_n) \in \Omega$ for every $t \in \partial_0 T_\varepsilon$. Differentiation under the sign of integration proves that g is analytic in Ω_ε'. If now $|z|$ is so small that $(t_1z_1, \cdots, t_nz_n) \in \Omega$ for every $t \in T_\varepsilon$, it follows from Theorem 2.2.1 that $f(z) = g(z)$. Since Ω_ε' is connected, we obtain that $f = g$ in Ω_ε'.

Now we have

$$(t_1 - 1)^{-1} \cdots (t_n - 1)^{-1} = \sum_\alpha t_1^{-\alpha_1 - 1} \cdots t_n^{-\alpha_n - 1}$$

with normal convergence when $t \in \partial_0 T_\varepsilon$. Since $(t_1 z_1, \cdots, t_n z_n)$ belongs to a compact set in Ω if z belongs to a compact set in Ω_ε' and $t \in \partial_0 T_\varepsilon$, we obtain with normal convergence in Ω_ε' that $f(z) = \Sigma f_\alpha(z)$, where

$$f_\alpha(z) = (2\pi i)^{-n} \int_{\partial_0 T_\varepsilon} f(t_1 z_1, \cdots, t_n z_n) t_1^{-\alpha_1 - 1} \cdots t_n^{-\alpha_n - 1}\, dt_1 \cdots dt_n$$

is analytic in Ω_ε'. As above, we find that $f_\alpha(z) = z^\alpha \partial^\alpha f(0)/\alpha!$ in a neighborhood of 0 and therefore in Ω_ε'. This completes the proof.

We shall say that a Reinhardt domain is logarithmically convex if it has the properties of a convergence domain which were described in Theorem 2.4.3. It is clear that the interior of the intersection of a family of logarithmically convex Reinhardt domains is also a logarithmically convex Reinhardt domain. To every open set Ω in $\mathbf{C}^n$ one can therefore find a smallest logarithmically convex Reinhardt domain containing it—possibly the whole of $\mathbf{C}^n$. From Theorems 2.4.3 and 2.4.5 we now obtain

Theorem 2.4.6. *Let Ω be a connected Reinhardt domain containing* 0 *and let $\tilde{\Omega}$ be the smallest logarithmically convex Reinhardt domain containing it. Every function in $A(\Omega)$ can then be extended to a function in $A(\tilde{\Omega})$.*

Example. *If* $\Omega = \{z \in \mathbf{C}^2, \max(|z_1|, |z_2|) < 1, \min(|z_1|, |z_2|) < \varepsilon\}$, *we have* $\tilde{\Omega} = \{z \in \mathbf{C}^2, \max(|z_1|, |z_2|, |z_1 z_2|/\varepsilon) < 1\}$.

2.5. Domains of holomorphy. In Theorems 2.3.2 and 2.4.6 we have found examples of open sets $\Omega \subsetneq \tilde{\Omega}$ such that every $u \in A(\Omega)$ can be extended to a function in $A(\tilde{\Omega})$. We shall now examine this phenomenon more closely.

Definition 2.5.1. *An open set $\Omega \subset \mathbf{C}^n$ is called a domain of holomorphy if there are no open sets Ω_1 and Ω_2 in $\mathbf{C}^n$ with the following properties:*

(a) $\varnothing \neq \Omega_1 \subset \Omega_2 \cap \Omega$.

(b) *Ω_2 is connected and not contained in Ω.*

(c) *For every $u \in A(\Omega)$ there is a function $u_2 \in A(\Omega_2)$ (necessarily uniquely determined) such that $u = u_2$ in Ω_1.*

Roughly speaking, the definition means that there is no part of the boundary across which every element in $A(\Omega)$ can be continued analytically. We shall see later that for every Ω there is a largest $\tilde{\Omega}$ to which all functions in $A(\Omega)$ can be continued analytically, and $\tilde{\Omega}$ is then a domain of holomorphy. However, this result is true only if we admit complex manifolds spread over $\mathbf{C}^n$. Since we only consider schlicht domains here, the discussion will be postponed to section 5.4.

Definition 2.5.2. *If K is a compact subset of Ω, we define the $A(\Omega)$-hull $\hat{K}_\Omega$ of K by*

$$\hat{K}_\Omega = \{z; z \in \Omega, |f(z)| \le \sup_K |f| \text{ if } f \in A(\Omega)\}. \tag{2.5.1}$$

If we choose $f(z) = \exp\langle z, \zeta\rangle$, we find that $\hat{K}_\Omega$ is contained in the convex hull of K, hence is bounded. It is also obvious that $\hat{K}_\Omega$ is closed in Ω, but in general $\hat{K}_\Omega$ need not be a compact subset of Ω (see Theorem 2.5.5 below).

If D is an open polydisc with center at 0, we set

$$\Delta_\Omega{}^D(z) = \sup\{r; \{z\} + rD \subset \Omega\}.$$

Lemma 2.5.3. *Let $f \in A(\Omega)$, assume that*

$$|f(z)| \le \Delta_\Omega{}^D(z), \qquad z \in K, \tag{2.5.2}$$

and let $\zeta \in \hat{K}_\Omega$. If $u \in A(\Omega)$, the power series expansion of u at ζ

$$\sum_\alpha (z - \zeta)^\alpha\, \partial^\alpha u(\zeta)/\alpha! \tag{2.5.3}$$

then converges when z belongs to the polydisc $\{\zeta\} + |f(\zeta)|D$.

Proof. Let $D = \{z; |z_j| < r_j, j = 1, \cdots, n\}$. If $0 < t < 1$, the set

$$\{z; |z_j - w_j| \le tr_j|f(w)|, j = 1, \cdots, n, \text{ for some } w \in K\}$$

is a compact subset of Ω, hence $|u(z)| \le M$ there for some M. From Cauchy's inequalities, we therefore obtain

$$|\partial^\alpha u(w)|t^{|\alpha|}r^\alpha|f(w)|^{|\alpha|}/\alpha! \le M, \qquad w \in K.$$

Since $f(w)^{|\alpha|}\,\partial^\alpha u(w)$ is analytic in Ω, the same estimate holds when $w \in \hat{K}_\Omega$. When $w = \zeta$, we conclude that (2.5.3) converges in the polydisc $\{\zeta\} + |f(\zeta)|D$.

Now let δ be an arbitrary continuous function in $\mathbf{C}^n$ such that $\delta > 0$ except at 0 and

$$\delta(tz) = |t|\delta(z), \qquad t \in \mathbf{C}, \quad z \in \mathbf{C}^n.$$

Set $\delta(z, \complement\Omega) = \inf_{w \in \complement\Omega} \delta(z - w)$. It is clear that $\delta(z, \complement\Omega)$ is a continuous function of z.

Theorem 2.5.4. *Let Ω be a domain of holomorphy. If $f \in A(\Omega)$ and*

$$|f(z)| \le \delta(z, \complement\Omega), \qquad z \in K,$$

where K is a compact subset of Ω, it follows that

$$|f(z)| \le \delta(z, \complement\Omega), \qquad z \in \hat{K}_\Omega.$$

In particular, when f is a constant we obtain

$$\inf_{z\in K, w\in\complement\Omega} \delta(z - w) = \inf_{z\in\hat{K}_\Omega, w\in\complement\Omega} \delta(z - w).$$

Proof. If $\{z; \delta(z) < 1\}$ is a polydisc D, the theorem follows from Lemma 2.5.3, since the power series expansions of all $u \in A(\Omega)$ at a point $\zeta \in \hat{K}_\Omega$ cannot converge in a fixed polydisc which is not contained in Ω, by the definition of a domain of holomorphy. Now we can write

$$\delta(z, \complement\Omega) = \sup\{r \in \mathbf{R}; z + aw \in \Omega \text{ if } w \in \mathbf{C}^n, \delta(w) \le 1, \text{ and } a \in \mathbf{C}, |a| < r\}$$
$$= \inf_{\delta(w)\le 1} \delta_w(z, \complement\Omega),$$

where

$$\delta_w(z, \complement\Omega) = \sup\{r \in \mathbf{R}; z + aw \in \Omega \text{ if } |a| < r\}.$$

It is therefore enough to prove the assertion for δ_w and we may of course assume that $w = (1, 0, \cdots)$. But if

$$D_k = \{z; |z_1| < 1, |z_2| < 1/k, \cdots, |z_n| < 1/k\},$$

then $\Delta_\Omega^{D_k}(z)$ increases to $\delta_w(z, \complement\Omega)$ when $k \to \infty$. If $\varepsilon > 0$, it follows from Dini's theorem that

$$|f(z)| \le \delta_w(z, \complement\Omega), \qquad z \in K,$$

implies

$$|f(z)| \le (1 + \varepsilon)\Delta_\Omega^{D_k}(z), \qquad z \in K,$$

if k is sufficiently large. Hence

$$|f(\zeta)| \le (1 + \varepsilon)\Delta_\Omega^{D_k}(\zeta) \le (1 + \varepsilon)\delta_w(\zeta, \complement\Omega), \qquad \zeta \in \hat{K}_\Omega,$$

according to the remark at the beginning of the proof. This proves the theorem.

We can now give a characterization of holomorphy domains.

Theorem 2.5.5. *If Ω is an open set in $\mathbf{C}^n$, the following conditions are equivalent:*

(i) *Ω is a domain of holomorphy.*

(ii) *If* $K \subset\subset \Omega$, *it follows that* $\hat{K}_\Omega \subset\subset \Omega$ *and that with the notations of Theorem* 2.5.4

$$\sup_{z \in K} |f(z)|/\delta(z, \complement\Omega) = \sup_{z \in \hat{K}_\Omega} |f(z)|/\delta(z, \complement\Omega), \qquad f \in A(\Omega).$$

(iii) *If* $K \subset\subset \Omega$, *we have* $\hat{K}_\Omega \subset\subset \Omega$.

(iv) *There exists a function* $f \in A(\Omega)$ *which cannot be continued analytically beyond* Ω, *that is, it is not possible to find* Ω_1 *and* Ω_2 *satisfying* (a) *and* (b) *in Definition* 2.5.1 *and* $f_2 \in A(\Omega_2)$ *so that* $f = f_2$ *in* Ω_1.

Proof. (i) ⇒ (ii) according to Theorem 2.5.4. The implications (ii) ⇒ (iii) and (iv) ⇒ (i) are trivial, so it remains to show only that (iii) ⇒ (iv). Let D be a polydisc with center at 0 and denote for $\zeta \in \Omega$ by D_ζ the largest polydisc of the form $\{\zeta\} + rD$ which is contained in Ω. Let M be a countable dense set in Ω. It is sufficient to construct $f \in A(\Omega)$ such that f cannot be continued to a neighborhood of $\bar{D}_\zeta$ for any $\zeta \in M$. To do so, we let $\zeta_1, \zeta_2, \cdots$ be a sequence of elements in M, containing every point in M infinitely many times. Let $K_1 \subset K_2 \subset \cdots$ be a sequence of compact subsets of Ω such that every compact subset of Ω belongs to some K_j. Since $\hat{K}_j \subset\subset \Omega$, we can find $z_j \in D_{\zeta_j}$ such that $z_j \notin \hat{K}_j$. Hence there is a function $f_j \in A(\Omega)$ such that $f_j(z_j) = 1$ but $\sup_{K_j}|f_j| < 1$. Replacing f_j by a high power of f_j, if necessary we may assume that

$$f_j(z_j) = 1, \qquad \sup_{K_j} |f_j| < 2^{-j}. \tag{2.5.4}$$

We can choose f_j so that f_j is not identically 1 in any of the components of Ω. Now form the infinite product

$$f = \prod_1^\infty (1 - f_j)^j.$$

Since $\Sigma j/2^j$ is convergent, the product converges uniformly on K_l for every l and therefore it defines a function $f \in A(\Omega)$ which is not identically 0 in any component of Ω. All derivatives of f of order $< j$ vanish at z_j. If $\zeta \in M$, we can thus for every integer N find points in D_ζ where all derivatives of f of order $\leq N$ are equal to 0. An analytic continuation of f to a neighborhood of $\bar{D}_\zeta$ would therefore have a zero of infinite order, hence be identically 0 in D_ζ, which is a contradiction. The proof is complete.

Corollary 2.5.6. *If* Ω *is convex in the geometric sense, then* Ω *is a domain of holomorphy.*

Corollary 2.5.7. *If Ω_α is a domain of holomorphy for every α in an index set A, then the interior Ω of $\cap_A \Omega_\alpha$ is a domain of holomorphy.*

Proof. Since $\hat{K}_\Omega \subset \hat{K}_{\Omega_\alpha}$ if $K \subset\subset \Omega$, the distance from $\hat{K}_\Omega$ to $\complement\Omega_\alpha$ is $\geq$ the distance from K to $\complement\Omega$ for every α. Hence condition (iii) in Theorem 2.5.5 is fulfilled.

Corollary 2.5.8. *Let Ω be a connected Reinhardt domain containing* 0. *Then the following conditions are equivalent:*

(i) *Ω is the domain of convergence of a power series.*

(ii) *Ω is a domain of holomorphy.*

(iii) $\Omega^* = \{\xi;\ \xi \in \mathbf{R}^n,\ (e^{\xi_1}, \cdots, e^{\xi_n}) \in \Omega\}$ *is an open convex set in* $\mathbf{R}^n$, *and if $\xi \in \Omega^*$ it follows that $\eta \in \Omega^*$ if $\eta_j \leq \xi_j$ for every j. Further, we have $z \in \Omega$ if and only if $|z_j| \leq e^{\xi_j}, j = 1, \cdots, n$, for some $\xi \in \Omega^*$.*

Proof. That (i) $\Rightarrow$ (iii) follows from Theorem 2.4.3. Further, (ii) $\Rightarrow$ (i) in view of Theorem 2.4.5, so we only have to prove that (iii) $\Rightarrow$ (ii). Let K be a compact set in Ω. We can find a finite set $k \subset \Omega$ such that

$$K \subset \bigcup_{\zeta \in k} \{z;\ |z_j| \leq |\zeta_j|,\ j = 1, \cdots, n\},$$

and no $|\zeta_j|$ vanishes when $\zeta \in k$. Assume now that $z \in \hat{K}_\Omega$ and that $z_1 \cdots z_j \neq 0$ while $z_{j+1}, \cdots, z_n$ all vanish. (We can reduce ourselves to this case by changing the notations for the coordinates.) Then we have for all α

$$|z_1^{\alpha_1} \cdots z_j^{\alpha_j}| \leq \sup_{\zeta \in k} |\zeta_1^{\alpha_1} \cdots \zeta_j^{\alpha_j}|,$$

that is, if $\lambda_i = \alpha_i/(\alpha_1 + \cdots + \alpha_j)$,

$$\sum_1^j \lambda_i \log|z_i| \leq \sup_{\zeta \in k} \sum_1^j \lambda_i \log|\zeta_i|.$$

Since λ_i are here arbitrary non-negative rationals with sum equal to 1, the estimate also holds for arbitrary non-negative reals λ_i. This means that the point $(\log|z_1|, \cdots, \log|z_j|) \in \mathbf{R}^j$ is in the convex hull of the set of all $(\eta_1, \cdots, \eta_j)$ such that $\eta_i \leq \log|\zeta_i|$, $i = 1, \cdots, j$, for some $\zeta \in k$. This proves that for some $\eta \in \Omega^*$ we have $|z_i| \leq e^{\eta_i}$, $i = 1, \cdots, j$, and therefore for all j. Hence $\hat{K}_\Omega \subset \Omega$, which proves the corollary.

We have thus proved that every function which is analytic in a connected Reinhardt domain containing 0 can be extended to an analytic function in a domain of holomorphy which is also a Reinhardt domain. The next theorem is analogous but more general.

Definition 2.5.9. *An open set $\Omega \subset \mathbf{C}^n$ is called a tube if there is an open set $\omega \subset \mathbf{R}^n$, called the base of Ω, such that $\Omega = \{z; \operatorname{Re} z \in \omega\}$.*

It is obvious that the convex hull $\operatorname{ch} \Omega$ is a tube with base $\operatorname{ch} \omega$.

Theorem 2.5.10. *If Ω is a connected tube, then every $u \in A(\Omega)$ can be extended to a function in $A(\operatorname{ch} \Omega)$.*

We shall deduce the theorem from Lemma 2.5.3 and

Lemma 2.5.11. *Let Ω be a tube whose base contains the convex hull of*

$$k = \{(x_1, 0, \cdots, 0), 0 \le x_1 \le 1\} \cup \{(0, x_2, \cdots, 0), 0 \le x_2 \le 1\}.$$

Set for $0 < \varepsilon < 1/2$,

$$K_\varepsilon = \{(x_1, x_2, 0, \cdots, 0); 0 \le x_1, 0 \le x_2, x_1 + x_2 \le 1,$$
$$x_1 + x_2 - \varepsilon(x_1{}^2 + x_2{}^2) \le 1 - \varepsilon\},$$

and

$$k_\varepsilon = \{x + iy; x \in k, y_1{}^2 + y_2{}^2 \le 1/\varepsilon, y_3 = \cdots = y_n = 0\}.$$

Then the $A(\Omega)$-hull of $k_\varepsilon + i\eta$ contains $K_\varepsilon + i\eta$ for every $\eta \in \mathbf{R}^n$.

Proof. We can take $\eta = 0$ in the proof. Now consider.

$$M_\varepsilon = \{(z_1, z_2, 0, \cdots, 0); \operatorname{Re} z_1 \ge 0, \operatorname{Re} z_2 \ge 0, \operatorname{Re}(z_1 + z_2) \le 1,$$
$$z_1 + z_2 - \varepsilon(z_1{}^2 + z_2{}^2) = 1 - \varepsilon\}.$$

With $z_j = x_j + iy_j$, we have in M_ε

$$x_1 + x_2 - \varepsilon(x_1{}^2 + x_2{}^2) + \varepsilon(y_1{}^2 + y_2{}^2) = 1 - \varepsilon, \tag{2.5.5}$$
$$x_1 \ge 0, x_2 \ge 0, x_1 + x_2 \le 1.$$

Hence $x_1{}^2 + x_2{}^2 \le 1$ and $y_1{}^2 + y_2{}^2 \le 1/\varepsilon$, so M_ε is compact. Since

$$\frac{\partial}{\partial z_1}(z_1 + z_2 - \varepsilon(z_1{}^2 + z_2{}^2)) = 1 - 2\varepsilon z_1 \ne 0 \quad \text{on } M_\varepsilon$$

the implicit function theorem shows that z_2 is locally an analytic function of z_1 on M_ε. Similarly, z_1 is locally an analytic function of z_2 on M_ε. Since $x_1 + x_2 < 1$ on M_ε except at the points $(1, 0, \cdots, 0)$ and $(0, 1, 0, \cdots, 0)$, the boundary of the surface M_ε belongs to k_ε and the maximum on M_ε of a function in $A(\Omega)$ is assumed in k_ε. Hence the $A(\Omega)$-hull of k_ε contains $\{(x_1, x_2, 0, \cdots, 0);\ x_1 \geq 0,\ x_2 \geq 0,\ x_1 + x_2 \leq 1,\ x_1 + x_2 - \varepsilon(x_1{}^2 + x_2{}^2) = 1 - \varepsilon\}$. Since $\lambda k_\varepsilon \subset k_\varepsilon$ when $0 \leq \lambda \leq 1$, the lemma follows.

Proof of Theorem 2.5.10. (a) We first assume that the base ω of Ω is starshaped with respect to the origin, that is, that $x \in \omega$, $0 \leq t \leq 1$ implies $tx \in \omega$. Note that every tube with a starshaped base is connected and that the intersection of two such tubes is starshaped and therefore connected. Hence there is a tube $\tilde{\Omega}$ with starshaped base $\tilde{\omega}$ such that every $u \in A(\Omega)$ can be continued analytically to $\tilde{\Omega}$, and $\tilde{\Omega}$ contains every starshaped tube with this property. In fact, we need only take the union of all such tubes. We have to prove that $\tilde{\omega}$ is convex. Let therefore ξ_1 and ξ_2 be two linearly independent elements in $\tilde{\omega}$. We may choose coordinates so that ξ_1 and ξ_2 are unit vectors along the x_1 and x_2 axes. Let k be defined as in Lemma 2.5.11 and choose $\delta > 0$ so that $\Delta_{\tilde{\Omega}}{}^D(z) > \delta$, $z \in k$ (for notations, see Lemma 2.5.3). Let

$$E = \{a;\ 0 \leq a \leq 1, \text{ such that } 0 \leq x_1, 0 \leq x_2, x_1 + x_2 \leq a \text{ implies } (x_1, x_2, 0, \cdots, 0) \in \tilde{\omega}\}.$$

It is obvious that E is open in $[0,1]$ and that $0 \in E$. If $a \in E$ and $0 < \varepsilon < 1/2$, it follows from Lemma 2.5.11 that the $A(\tilde{\Omega})$-hull of $k + iK$ for some compact set $K \in \mathbf{R}^n$ contains

$$L_{a,\varepsilon} = \{(x_1, x_2, 0, \cdots, 0);\ 0 \leq x_1, 0 \leq x_2, x_1 + x_2 \leq (1 - \varepsilon)a\}.$$

Hence the power series expansion of any $\tilde{f} \in A(\tilde{\Omega})$ at a point ζ with $\operatorname{Re}\zeta \in L_{a,\varepsilon}$ converges in $\{\zeta\} + \delta D$. From this we conclude that E is closed, hence $E = [0,1]$, which proves the convexity of $\tilde{\omega}$.

(b) Now let ω be an arbitrary connected open set in $\mathbf{R}^n$. Let $0 \in \omega$ and denote by $\tilde{\Omega}$ the largest tube whose base is starshaped with respect to the origin such that, to every $f \in A(\Omega)$, one can find $\tilde{f} \in A(\tilde{\Omega})$ so that $f = \tilde{f}$ in a neighborhood of 0. According to (a), $\tilde{\Omega}$ must be convex. Now assume that $\tilde{\Omega}$ does not contain all of Ω. Then we can find a point $x_0 \in \omega$ so that $x_0 \notin \tilde{\omega}$ and, since ω is connected, we can join x_0 to 0 with a polygon in ω. Let x_1 be its last intersection with $\partial\tilde{\omega}$. Then x_1 is connected to 0 with a polygon which apart from x_1 belongs to $\omega \cap \tilde{\omega}$. If ω_1 is a convex neighborhood of x_1 which is contained in ω, the function $f' = \tilde{f}$ in $\tilde{\omega} + i\mathbf{R}^n$, $f' = f$ in $\omega_1 + i\mathbf{R}^n$ is uniquely defined and analytic in the tube with base $\tilde{\omega} \cup \omega_1$, which is starshaped with respect to x_1. According to (a) the function f' can therefore be continued to the tube with base $\operatorname{ch}(\tilde{\omega} \cup \omega_1)$, which is starshaped with respect to 0.

also. This contradicts the definition of $\tilde{\Omega}$. Hence $\tilde{\Omega} \supset \Omega$, and we obtain $\tilde{f} = f$ in the whole of Ω by the uniqueness of analytic continuation.

Example. If P is a polynomial (or entire function), the components of the interior of $\{\xi;\, P(\xi + i\eta) \neq 0 \text{ for all } \eta \in \mathbf{R}^n\}$ are all convex. This follows from Theorem 2.5.10 applied to $1/P$. In particular, this gives interesting information concerning hyperbolic polynomials (cf. Hörmander [2], Chapter V).

Corollary 2.5.12 (*to Theorem* 2.5.10). *A tube is a domain of holomorphy if and only if every component is convex.*

We finally list some simple constructions which lead to domains of holomorphy.

Theorem 2.5.13. *Let Ω be a domain of holomorphy and $f_1, \cdots, f_N \in A(\Omega)$. Then $\Omega_f = \{z;\, z \in \Omega, |f_j(z)| < 1, j = 1, \cdots, N\}$ is a domain of holomorphy.*

Proof. Let K be a compact set in Ω_f. Choose $r < 1$ so that $|f_j| \le r$ in K when $j = 1, \cdots, N$. Then this inequality is also valid in $\hat{K}_\Omega$, which proves that $\hat{K}_\Omega \subset \Omega_f$. Now $\hat{K}_\Omega$ is compact because Ω is a domain of holomorphy, and since $\hat{K}_{\Omega_f} \subset \hat{K}_\Omega$ it follows that $\hat{K}_{\Omega_f}$ is compact.

A more general form of the same statement is the following:

Theorem 2.5.14. *Let Ω and Ω' be holomorphy domains in $\mathbf{C}^n$ and in $\mathbf{C}^m$, respectively, and let u be an analytic map of Ω into $\mathbf{C}^m$. Then*

$$\Omega_u = \{z;\, z \in \Omega, u(z) \in \Omega'\}$$

is a domain of holomorphy.

Proof. Let K be a compact set in Ω_u. Since $\hat{K}_{\Omega_u} \subset \hat{K}_\Omega \subset\subset \Omega$, it is sufficient to prove that $\hat{K}_{\Omega_u}$ is closed in Ω. Now $u(K)$ is a compact subset of Ω' since u is continuous, so the $A(\Omega')$-hull K' of $u(K)$ is a compact subset of Ω'. If $f \in A(\Omega')$ and $\zeta \in \hat{K}_{\Omega_u}$, we have

$$|f(u(\zeta))| \le \sup_{z \in K} |f(u(z))| = \sup_{w \in u(K)} |f(w)|,$$

which means that $u(\zeta) \in K'$. Hence every point in the closure of $\hat{K}_{\Omega_u}$ in Ω is mapped by u into K' and therefore belongs to Ω_u. The proof is complete.

Remark. The proof does not use the full force of the hypothesis that Ω is a domain of holomorphy; we only needed that $\hat{K}_{\Omega_u} \subset\subset \Omega$ for every

$K \subset\subset \Omega_u$. If $\Omega_u \subset\subset \Omega$, we can therefore drop the assumption that Ω is a domain of holomorphy.

2.6. Pseudoconvexity and plurisubharmonicity. So far we have only used Theorem 2.5.4 when f is a constant, but we shall now extract further information. Let z_0 belong to a domain of holomorphy Ω and let $w \in \mathbf{C}^n$. Choose r so small that

$$D = \{z_0 + \tau w;\ \tau \in \mathbf{C},\ |\tau| \leq r\} \subset \Omega,$$

and let $f(\tau)$ be an analytic polynomial, such that

$$-\log \delta(z_0 + \tau w, \complement \Omega) \leq \operatorname{Re} f(\tau), \qquad |\tau| = r.$$

If we choose an analytic polynomial F in $\mathbf{C}^n$ so that $F(z_0 + \tau w) = f(\tau)$, our hypothesis can be written

$$|e^{-F(z)}| \leq \delta(z, \complement \Omega), \qquad z \in \partial D.$$

Since the $A(\Omega)$-hull of ∂D contains D by the maximum principle, it follows from Theorem 2.5.4 that $|e^{-F(z)}| \leq \delta(z, \complement \Omega)$, $z \in D$, that is,

$$-\log \delta(z_0 + \tau w, \complement \Omega) \leq \operatorname{Re} f(\tau), \qquad |\tau| \leq r.$$

The same conclusion is obvious if $w = 0$. Hence $-\log \delta(z + \tau w, \complement \Omega)$ is for fixed $z \in \mathbf{C}^n$ and $w \in \mathbf{C}^n$ a subharmonic function of τ where it is defined. (See condition (i) in Theorem 1.6.3). We introduce a name for such functions:

Definition 2.6.1. *A function u defined in an open set $\Omega \subset \mathbf{C}^n$ with values in $[-\infty, +\infty)$ is called plurisubharmonic if*

(a) *u is semicontinuous from above.*
(b) *For arbitrary z and $w \in \mathbf{C}^n$, the function $\tau \to u(z + \tau w)$ is subharmonic in the part of $\mathbf{C}$ where it is defined.*

We shall denote the set of all such functions by $P(\Omega)$.

Example. If f is analytic, then $\log|f|$ is plurisubharmonic.

Before proceeding further we shall list the properties of plurisubharmonic functions which we shall need, and which are not immediate consequences of their analogues for subharmonic functions.

Theorem 2.6.2. *A function $u \in C^2(\Omega)$ is plurisubharmonic if and only if*

$$\sum_{j,k=1}^{n} \partial^2 u(z)/\partial z_j \partial \bar{z}_k\, w_j \bar{w}_k \geq 0, \qquad z \in \Omega,\ w \in \mathbf{C}^n. \tag{2.6.1}$$

Proof. We only have to apply Theorem 1.6.10 to the functions

$$\tau \to u(z + \tau w).$$

Theorem 2.6.3. *Let $0 \le \varphi \in C_0^\infty(\mathbf{C}^n)$ be equal to 0 when $|z| > 1$, let φ depend only on $|z_1|, \cdots, |z_n|$, and assume that $\int \varphi(z)\, d\lambda(z) = 1$ where $d\lambda$ is the Lebesgue measure. If u is plurisubharmonic in Ω, it follows that*

$$u_\varepsilon(z) = \int u(z - \varepsilon\zeta)\varphi(\zeta)\, d\lambda(\zeta)$$

is plurisubharmonic, that $u_\varepsilon \in C^\infty$ where $d(z, \complement\Omega) > \varepsilon$, and that $u_\varepsilon \searrow u$ when $\varepsilon \searrow 0$. (We assume that $u \not\equiv -\infty$.)

Proof. That u_ε decreases when $\varepsilon \searrow 0$ was proved in the case $n = 1$ in the proof of Theorem 1.6.11. Iteration of this result shows that u_ε is decreasing also if $n > 1$, and from the case $n = 1$ we also immediately find that $u \le u_\varepsilon$. Since $\overline{\lim}_{\varepsilon \to 0} u_\varepsilon \le u$ in view of the upper semi-continuity of u, we conclude that $u_\varepsilon \searrow u$ when $\varepsilon \searrow 0$. That u_ε is plurisubharmonic follows immediately from Theorem 1.6.10.

Conversely, Theorem 1.6.2 shows immediately that the limit of a decreasing sequence of plurisubharmonic functions is plurisubharmonic.

Theorem 2.6.4. *Let $\Omega \subset \mathbf{C}^n$ and $\Omega' \subset \mathbf{C}^m$, let f be an analytic map of Ω into Ω', and let $u \in P(\Omega')$. Then $f^*u \in P(\Omega)$.*

Proof. First assume that $u \in C^2(\Omega')$. Then we have

$$\sum_{j,k=1}^{n} \partial^2 u(f(z))/\partial z_j \partial \bar{z}_k\, w_j \bar{w}_k = \sum_{j,k=1}^{m} \partial^2 u/\partial f_j \partial \bar{f}_k\, g_j \bar{g}_k \ge 0,$$

where we have written $g_j = \Sigma_1^n w_i \partial f_j/\partial z_i$. Hence $f^* u \in P(\Omega)$. For a general $u \in P(\Omega')$, we only have to use Theorem 2.6.3 to choose a sequence of C^∞ plurisubharmonic functions which decrease toward u and use the remark preceding the theorem.

Note that Theorem 2.6.4 strengthens condition (b) in Definition 2.6.1 even when $n = 1$.

The result obtained at the beginning of this section can now be stated as follows:

Theorem 2.6.5. *If Ω is a domain of holomorphy and $\delta(z, \complement\Omega)$ is defined as in Theorem 2.5.4, then $-\log \delta(z, \complement\Omega)$ is plurisubharmonic and continuous.*

In Chapter IV we shall prove that the converse of this result is true.

However, we can already at this point discuss equivalent forms of the condition just obtained.

Definition 2.6.6. *If K is a compact subset of the open set $\Omega \subset \mathbf{C}^n$ we define the $P(\Omega)$-hull $\hat{K}_\Omega^P$ of K by*

$$\hat{K}_\Omega^P = \{z\,;\, z \in \Omega,\, u(z) \le \sup_K u \text{ for all } u \in P(\Omega)\}.$$

It is clear that the $P(\Omega)$-hull of K is contained in the $A(\Omega)$-hull of K.

Theorem 2.6.7. *If Ω is an open set in $\mathbf{C}^n$, the following conditions are equivalent:*

(i) $-\log \delta(z, \complement\Omega)$ *is plurisubharmonic in Ω if δ is defined as in Theorem* 2.5.4.

(ii) *There exists a continuous plurisubharmonic function u in Ω such that*

$$\Omega_c = \{z\,;\, z \in \Omega,\, u(z) < c\} \subset\subset \Omega$$

for every $c \in \mathbf{R}$.

(iii) $\hat{K}_\Omega^P \subset\subset \Omega$ *if* $K \subset\subset \Omega$.

Proof. If (i) is fulfilled, we only have to set $u(z) = |z|^2 - \log \delta(z, \complement\Omega)$ to get a function satisfying (ii). That (ii) implies (iii) is obvious, so we need only prove that (iii) implies (i). Let $z_0 \in \Omega$, $0 \ne w \in \mathbf{C}^n$, and choose $r > 0$ so that

$$D = \{z_0 + \tau w\,;\, |\tau| \le r\} \subset \Omega.$$

Let $f(\tau)$ be an analytic polynomial such that

$$-\log \delta(z_0 + \tau w, \complement\Omega) \le \operatorname{Re} f(\tau), \qquad |\tau| = r,$$

that is,

$$\delta(z_0 + \tau w, \complement\Omega) \ge |e^{-f(\tau)}|, \qquad |\tau| = r. \tag{2.6.2}$$

We want to prove the same inequality when $|\tau| \le r$. To do so, we take any vector $a \in \mathbf{C}^n$ with $\delta(a) < 1$ and consider for $0 \le \lambda \le 1$ the mapping

$$\tau \to z_0 + \tau w + \lambda a e^{-f(\tau)}, \qquad |\tau| \le r.$$

We denote its range by D_λ; clearly $D_0 = D$. Put

$$\Lambda = \{\lambda\,;\, 0 \le \lambda \le 1,\, D_\lambda \subset \Omega\}.$$

It is obvious that Λ is an open subset of $[0,1]$ and to prove that Λ is equal to the whole interval we shall prove that Λ is closed. Let K be the compact set

$$K = \{z_0 + \tau w + \lambda a e^{-f(\tau)}, |\tau| = r, 0 \leq \lambda \leq 1\},$$

which is contained in Ω by (2.6.2). If $u \in P(\Omega)$ and $\lambda \in \Lambda$, then

$$\tau \to u(z_0 + \tau w + \lambda a e^{-f(\tau)})$$

is subharmonic in a neighborhood of the disc $|\tau| \leq r$, which proves that

$$u(z_0 + \tau w + \lambda a e^{-f(\tau)}) \leq \sup_K u \quad \text{if} \quad |\tau| \leq r.$$

Hence $D_\lambda \subset \hat{K}_\Omega{}^P$ for every $\lambda \in \Lambda$, and this implies that Λ is closed, for $\hat{K}_\Omega{}^P$ is relatively compact in Ω by (iii). Thus $D_1 \subset \Omega$, that is,

$$z_0 + \tau w + a e^{-f(\tau)} \in \Omega \quad \text{if } \delta(a) < 1 \text{ and } |\tau| \leq r,$$

so that $|\delta(z_0 + \tau w, \complement\,\Omega)| \geq |e^{-f(\tau)}|$ if $|\tau| \leq r$, or

$$-\log \delta(z_0 + \tau w, \complement\,\Omega) \leq \operatorname{Re} f(\tau), \qquad |\tau| \leq r.$$

This proves (i).

Definition 2.6.8. *The open set $\Omega \subset \mathbf{C}^n$ is called pseudoconvex if the equivalent conditions in Theorem 2.6.7 are fulfilled.*

Since the supremum of a family of plurisubharmonic functions is plurisubharmonic if it is continuous, we obtain from condition (i) in Theorem 2.6.7:

Theorem 2.6.9. *If Ω_α is a pseudoconvex open set for every α in an index set A, then the interior Ω of $\cap_{\alpha \in A} \Omega_\alpha$ is also pseudoconvex.*

The corresponding statement for holomorphy domains was given in Corollary 2.5.7. However, the next theorem is by no means obvious for domains of holomorphy and is, in fact, in that case essentially equivalent to the identity of holomorphy domains and pseudoconvex domains which we shall prove in Chapter IV.

Theorem 2.6.10. *Let Ω be an open set in $\mathbf{C}^n$. If to every point in $\bar{\Omega}$ there is a neighborhood ω such that $\omega \cap \Omega$ is pseudoconvex, then Ω is pseudoconvex.*

The condition in the theorem is of course only a restriction on $\partial\Omega$. Loosely stated, the theorem means that pseudoconvexity is a local property of the boundary.

Proof of Theorem 2.6.10. Let $z_0 \in \partial\Omega$ and choose a neighborhood ω of z_0 according to the hypothesis. Then $\delta(z, \complement\Omega) = \delta(z, \complement(\Omega \cap \omega))$ for all z sufficiently close to z_0. Hence $-\log \delta(z, \complement\Omega)$ is plurisubharmonic in a neighborhood of every point on $\partial\Omega$, so there is a closed set $F \subset \Omega$ such that $-\log \delta(z, \complement\Omega)$ is plurisubharmonic in $\Omega \setminus F$. Now take a continuous function $\varphi \in P(\mathbf{C}^n)$ (for example, a convex increasing function of $|z|^2$) such that $\varphi(z) > -\log \delta(z, \complement\Omega)$ when $z \in F$, and $\varphi(z) \to \infty$ when $|z| \to \infty$. Then $u(z) = \sup(\varphi(z), -\log \delta(z, \complement\Omega))$ is in $P(\Omega)$, for $u = \varphi$ in a neighborhood of F and the supremum of two plurisubharmonic functions is plurisubharmonic. It is clear that u satisfies condition (ii) in Theorem 2.6.7, which proves that Ω is pseudoconvex.

For reference in Chapter IV we give a property of the $P(\Omega)$-hull which seems much stronger than that in Definition 2.6.6.

Theorem 2.6.11. *Let Ω be a pseudoconvex open set in $\mathbf{C}^n$, let K be a compact subset of Ω, and ω an open neighborhood of $\hat{K}_\Omega^P$. Then there exists a function $u \in C^\infty(\Omega)$ such that*

(a) *u is strictly plurisubharmonic, that is, the hermitian form in* (2.6.1) *is strictly positive definite for every $z \in \Omega$.*

(b) *$u < 0$ in K but $u > 0$ in $\Omega \cap \complement\omega$.*

(c) *$\{z; z \in \Omega, u(z) < c\} \subset\subset \Omega$ for every $c \in \mathbf{R}$.*

Proof. We shall first construct a continuous function $v \in P(\Omega)$ satisfying (b) and (c). To do so we choose a function u_0 with the properties listed in (ii) of Theorem 2.6.7. Adding a constant to u_0, if necessary, we may assume that $u_0 < 0$ in K. Set

$$K' = \{z; z \in \Omega, u_0(z) \le 2\}, L = \{z; z \in \Omega \cap \complement\omega, u_0(z) \le 0\}.$$

These sets are both compact. For every $z \in L$ we can choose a function $w \in P(\Omega)$ such that $w(z) > 0$ but $w < 0$ in K. By means of a regularization as in Theorem 2.6.3 we obtain a continuous plurisubharmonic function w_1 in a neighborhood of K', with $w_1 < 0$ in K and $w_1 > 0$ in a neighborhood of z. Since L is compact, we can now use the Borel–Lebesgue lemma and the fact that the supremum of a finite family of plurisubharmonic functions is plurisubharmonic to construct a continuous plurisubharmonic function w_2 in a neighborhood of K' such that $w_2 > 0$ in a neighborhood of L and $w_2 < 0$ in K. Let C be the maximum of w_2 in K' and set for $z \in \Omega$,

$$v(z) = \sup(w_2(z), Cu_0(z)) \quad \text{if } u_0(z) < 2, \qquad v(z) = Cu_0(z) \quad \text{if } u_0(z) > 1.$$

The two definitions agree when $1 < u_0(z) < 2$, so v is a continuous plurisubharmonic function in Ω, which obviously satisfies (b) and (c).

Let

$$\Omega_c = \{z; z \in \Omega, v(z) < c\}.$$

If with the notations of Theorem 2.6.3 we set

$$v_j(z) = \int_{\Omega_{j+1}} v(\zeta)\varphi((z - \zeta)/\varepsilon)\varepsilon^{-2n}\, d\lambda(\zeta) + \varepsilon|z|^2 \quad (j = 0,1, \cdots)$$

and ε is chosen sufficiently small (depending on j), we obtain a function $v_j \in C^\infty(\mathbf{C}^n)$ which is $> v$ and strictly plurisubharmonic in a neighborhood of $\bar{\Omega}_j$. By suitable choice of ε we can also achieve that $v_0 < 0$ and $v_1 < 0$ in K and $v_j < v + 1$ in Ω_j for $j = 1, 2, \cdots$. Now take a convex function $\chi \in C^\infty(\mathbf{R})$ such that $\chi(t) = 0$ when $t < 0$ and $\chi'(t) > 0$ when $t > 0$. Then $\chi(v_j + 1 - j)$ is strictly plurisubharmonic in a neighborhood of $\bar{\Omega}_j \setminus \Omega_{j-1}$. We can therefore choose successively positive numbers $a_1, a_2, \cdots$ so that

$$u_m = v_0 + \sum_1^m a_j \chi(v_j + 1 - j)$$

is $> v$ and strictly plurisubharmonic in a neighborhood of $\bar{\Omega}_m$. We have $u_m = u_l$ in Ω_j if l and m are $> j$, so $u = \lim u_m$ exists and is a strictly plurisubharmonic C^∞ function in Ω. Since $u = v_0 < 0$ in K and $u > v$ in Ω, the properties (a), (b), (c) follow.

If Theorem 2.6.11 is applied to $\omega = \Omega \setminus \{x\}$, where $x \notin \hat{K}_\Omega^P$, it follows that we can restrict u to $P(\Omega) \cap C^\infty(\Omega)$ in Definition 2.6.6. Hence $\hat{K}_\Omega^P$ is closed and therefore compact if K is a compact subset of Ω and Ω is pseudoconvex.

We shall now examine when an open set with a C^2 boundary is pseudoconvex.

Theorem 2.6.12. *Let $\Omega \subset \mathbf{C}^n$ be an open set with a C^2 boundary; let $\Omega = \{z; \rho(z) < 0\}$ where ρ is in C^2 in a neighborhood of $\bar{\Omega}$ and* grad $\rho \neq 0$ *on $\partial\Omega$. Then Ω is pseudoconvex if and only if*

$$(2.6.3)\quad \sum_{j,k=1}^n \partial^2\rho/\partial z_j \partial \bar{z}_k\, w_j \bar{w}_k \geq 0 \quad \text{when } z \in \partial\Omega \text{ and} \sum_1^n \partial\rho/\partial z_j\, w_j = 0.$$

Condition (2.6.3) is called the *Levi condition*; $\partial\Omega$ is also said to be *pseudoconvex* if (2.6.3) holds.

Proof. If ρ_1 is another function satisfying the hypotheses in the theorem, then $\rho_1 = h\rho$ with $h > 0$ in a neighborhood of $\bar{\Omega}$. Hence

$$\sum_{j,k=1}^n \partial^2\rho_1/\partial z_j \partial \bar{z}_k\, w_j \bar{w}_k = h \sum_{j,k=1}^n \partial^2\rho/\partial z_j \partial \bar{z}_k\, w_j \bar{w}_k \quad \text{if } \rho = \sum_1^n \partial\rho/\partial z_j\, w_j = 0;$$

which proves that (2.6.3) is independent of the choice of ρ. To prove that there is a function satisfying (2.6.3) if Ω is pseudoconvex, we set

$$\rho(z) = -\delta(z, \complement\Omega) \quad \text{in } \Omega, \qquad \rho(z) = \delta(z,\Omega) \quad \text{in } \complement\Omega$$

where δ is, for example, the Euclidean metric. Then $\rho \in C^2$ near the boundary of Ω. (This follows from the implicit function theorem.) At points in Ω sufficiently close to $\partial\Omega$, the plurisubharmonicity of $-\log\delta$ means that

$$\sum_{j,k=1}^{n} (-\delta^{-1}\,\partial^2\delta/\partial z_j\partial\bar{z}_k + \delta^{-2}\,\partial\delta/\partial z_j\partial\delta/\partial\bar{z}_k)\,w_j\bar{w}_k \geq 0.$$

Hence

$$\sum_{j,k=1}^{n} \partial^2\rho/\partial z_j\partial\bar{z}_k\,w_j\bar{w}_k \geq 0 \quad \text{if} \sum_{1}^{n} \partial\rho/\partial z_j\,w_j = 0.$$

A passage to the limit shows that this is also true on $\partial\Omega$.

In proving the converse we may by the first part of the proof assume that (2.6.3) is satisfied with ρ defined as above. Assume now that with $w \in \mathbf{C}^n$

$$c = \frac{\partial^2}{\partial\tau\partial\bar{\tau}}\log\delta(z + \tau w, \complement\Omega) > 0 \quad \text{when } \tau = 0,$$

for some z so close to $\partial\Omega$ that $\delta \in C^2$ at z. Then we have by Taylor's formula

$$\log\delta(z + \tau w, \complement\Omega) = \log\delta(z, \complement\Omega) + \operatorname{Re}(A\tau + B\tau^2) + c|\tau|^2 + o(|\tau|^2), \qquad \tau \to 0.$$

Here A and B are constants. Now choose $a \in \mathbf{C}^n$ with $\delta(a) = \delta(z, \complement\Omega)$ so that $z + a \in \partial\Omega$, and set

$$z(\tau) = z + \tau w + a\exp(A\tau + B\tau^2).$$

Then we have

$$\begin{aligned}\delta(z(\tau), \complement\Omega) &\geq \delta(z + \tau w, \complement\Omega) - \delta(a)|\exp(A\tau + B\tau^2)| \\ &\geq \delta(a)(e^{c|\tau|^2/2} - 1)|e^{A\tau + B\tau^2}|\end{aligned}$$

when $|\tau|$ is sufficiently small. Since $\delta(z(0), \complement\Omega) = 0$, we conclude that $(\partial/\partial\tau)\delta(z(\tau), \complement\Omega) = 0$ and that $(\partial^2/\partial\tau\partial\bar{\tau})\delta(z(\tau), \complement\Omega) > 0$ when $\tau = 0$. With the notation ρ used above, this means, since $z(\tau)$ is an analytic function of τ, that

$$\sum_{1}^{n} \partial\rho/\partial z_j\,z_j'(0) = 0, \qquad \sum_{j,k=1}^{n} \partial^2\rho/\partial z_j\partial\bar{z}_k\,z_j'(0)\,\overline{z_k'(0)} < 0.$$

This contradicts (2.6.3) at $z(0)$, which completes the proof. (Note the analogy of this proof with that of Theorem 2.6.7.)

The necessity of (2.6.3) is also easily proved directly. In fact, when (2.6.3) is violated, one can even prove a local extension theorem analogous to Theorem 2.3.2'.

Theorem 2.6.13. *Let ρ be a C^4 function in a neighborhood ω of z_0 such that $\rho(z_0) = 0$ but* $\operatorname{grad}\rho(z_0) \neq 0$. *Further, assume that*

(2.6.4)
$$\sum_{j,k=1}^{n} \partial^2\rho(z_0)/\partial z_j\partial\bar{z}_k\, w_j\bar{w}_k < 0 \quad \textit{for some } w \in \mathbf{C}^n \textit{ with } \sum_1^n w_j\partial\rho(z_0)/\partial z_j = 0.$$

Then there exists a neighborhood $\omega' \subset \omega$ of z_0 such that, for every $u \in C^4(\omega')$ satisfying the tangential Cauchy–Riemann equations, $\bar\partial u \wedge \bar\partial\rho = 0$ on $\{z; z \in \omega', \rho(z) = 0\}$, one can find $U \in C^1(\omega')$ so that $U = u$ when $\rho = 0$, and $\bar\partial U = 0$ in $\omega_+' = \{z; z \in \omega', \rho(z) \geq 0\}$.

Proof. After a linear change of coordinates we may assume that $z_0 = 0$ and that

$$\rho(x) = x_{2n} + A(x) + O(|x|^3)$$

where A is a quadratic form. Taylor's formula gives

$$A(x) = \sum_{j,k=1}^{n} z_j\bar{z}_k\, \partial^2\rho(0)/\partial z_j\partial\bar{z}_k + \operatorname{Re}\sum_{j,k=1}^{n} z_j z_k\, \partial^2\rho(0)/\partial z_j\partial z_k.$$

If we make the analytic change of variables

$$z_j' = z_j \quad \text{if } j \leq n-1,\ z_n' = z_n + i\sum_{j,k=1}^{n} z_j z_k\, \partial^2\rho(0)/\partial z_j\partial z_k,$$

the Taylor expansion of ρ assumes the simpler form

$$\rho = \operatorname{Im} z_n' + \sum_{j,k=1}^{n} z_j'\overline{z_k'}\, \partial^2\rho(0)/\partial z_j\partial\bar{z}_k + O(|z'|^3).$$

To simplify notations we may therefore without restriction assume that the original coordinates were chosen so that

$$\rho = \operatorname{Im} z_n + \sum_{j,k=1}^{n} A_{jk} z_j\bar{z}_k + O(|z|^3),$$

where (A_{jk}) is a hermitian symmetric matrix. The hypothesis (2.6.4) means that the form

$$\sum_{j,k=1}^{n-1} A_{jk}\, z_j \bar{z}_k$$

is not positive definite. By a linear change of coordinates we may achieve that $A_{11} < 0$. Since

$$\rho(z_1, 0, \cdots, 0) = A_{11}|z_1|^2 + O(|z_1|^3)$$

we can then choose first $\delta > 0$ and then $\varepsilon > 0$ so that $\partial^2\rho/\partial z_1 \partial \bar{z}_1 < 0$ in

$$\omega' = \{z\,;\, |z_1| < \delta, |z_2| + \cdots + |z_n| < \varepsilon\} \subset \omega,$$

and $\rho < 0$ on the part of the boundary where $|z_1| = \delta$. For fixed $z_2, \cdots, z_n$ with $|z_2| + \cdots + |z_n| < \varepsilon$, the set of all z_1 with $|z_1| < \delta$ where $\rho < 0$ is then connected, because ρ cannot have a local minimum as a function of z_1. The proof of Theorem 2.3.1 therefore applies without change to prove that for every form $f \in C^k_{(0,1)}(\omega')$, $k > 0$, satisfying the equation $\bar{\partial} f = 0$ and vanishing outside ω_+' there exists a solution $u \in C^k(\omega')$ of the equation $\bar{\partial} u = f$ such that u vanishes outside ω_+'. But then the assertion of the theorem follows by repetition of the proof of Theorem 2.3.2′. The details may be left as an exercise for the reader.

If on a smooth surface in $\mathbf{C}^n$ we are given a function u satisfying the tangential Cauchy–Riemann equations, it is thus possible to extend u analytically to at least one side of the hypersurface in a neighborhood of every point where the Levi form (2.6.3) does not vanish identically. In that case no extension need exist, however. For example, any function of x_{2n-1} satisfies the tangential Cauchy–Riemann equations in the hyperplane $x_{2n} = 0$, but obviously there is no analytic extension in general.

2.7. Runge domains. We shall here prove an approximation theorem extending Corollary 1.3.2 and in the same context an existence theorem for the $\bar{\partial}$ operator. The elementary methods which we use here do not allow us to study general holomorphy domains, so we shall return to these questions in Chapter IV.

Definition 2.7.1. *A domain of holomorphy $\Omega \subset \mathbf{C}^n$ is called a Runge domain if polynomials are dense in $A(\Omega)$, that is, if every $f \in A(\Omega)$ can be uniformly approximated on an arbitrary compact set in Ω by analytic polynomials.*

Since power series expansions show that polynomials are dense in $A(\mathbf{C}^n)$, we might as well have considered arbitrary entire functions instead of polynomials in the definition. In what follows we set for compact sets K (cf. Definition 2.5.2)

$$\tilde{K} = \hat{K}_{\mathbf{C}^n} = \{z\,;\, z \in \mathbf{C}^n, |P(z)| \le \sup_K |P| \text{ for all polynomials } P\}.$$

Definition 2.7.2. *Compact sets K with $K = \tilde{K}$ are called polynomially convex.*

We shall now prove (cf. Theorem 1.3.4)

Theorem 2.7.3. *The following conditions on a domain of holomorphy $\Omega \subset \mathbf{C}^n$ are equivalent:*

(i) *Ω is a Runge domain.*
(ii) *For every compact set $K \subset \Omega$ we have $\tilde{K} = \hat{K}_\Omega$.*
(iii) *For every compact set $K \subset \Omega$ we have $\tilde{K} \cap \Omega = \hat{K}_\Omega$.*
(iv) *For every compact set $K \subset \Omega$ we have $\tilde{K} \cap \Omega \subset\subset \Omega$.*

Proof. That (i) $\Rightarrow$ (iii) follows from the definition of $\hat{K}_\Omega$ and the definition of a Runge domain. The implication (ii) $\Rightarrow$ (iii) is trivial and (iii) $\Rightarrow$ (iv) since Ω is a domain of holomorphy. The remaining implications (iv) $\Rightarrow$ (ii) $\Rightarrow$ (i) will be proved later on in this section after we have studied the equation $\bar{\partial} u = f$ in neighborhoods of polynomially convex sets.

We shall say that a compact set K has the *Cousin property* if for every $f \in C^\infty_{(p,q+1)}$ with $\bar{\partial} f = 0$ in a neighborhood of $K (p, q \ge 0)$ the equation $\bar{\partial} u = f$ has a solution $u \in C^\infty_{(p,q)}$ in some neighborhood of K. Our aim is to prove that all polynomially convex sets have the Cousin property. First we approximate by sets of a simpler type.

Lemma 2.7.4. *Let K be a polynomially convex compact set and let ω be a neighborhood of K. Then one can find polynomials $P_1, \cdots, P_m$ such that*

$$K \subset \{z\,;\, |P_j(z)| \le 1, j = 1, \cdots, m\} = L \subset \omega.$$

L is called a polynomial polyhedron, and is obviously polynomially convex.

Proof. Let $P_j(z) = a_j z_j, j = 1, \cdots, n$, where $a_j > 0$ are chosen so small that $|P_j| \le 1$ in K. For every $z \in \complement\omega$ with $|a_j z_j| \le 1$ for $j = 1, \cdots, n$, we can find a polynomial P so that $|P(z)| > 1$ but $\sup_K |P| \le 1$. Application of the Borel–Lebesgue lemma now shows that $P_{n+1}, \cdots, P_m$ can be chosen so that the statement of the lemma is valid.

The following fundamental lemma is essentially due to Oka [1].

Lemma 2.7.5. *Let K be a compact set in $\mathbf{C}^n$, let D be the closed unit disc in $\mathbf{C}$, and let P be a polynomial in $\mathbf{C}^n$. Denote by μ the Oka map*

$$\mathbf{C}^n \ni z \to (z, P(z)) \in \mathbf{C}^{n+1},$$

and set $K_P = \{z; z \in K, |P(z)| \leq 1\}$. Assume that the set $K \times D$ in $\mathbf{C}^{n+1}$ has the Cousin property. Then

(a) *K_P has the Cousin property.*
(b) *For every $f \in C^\infty_{(p,q)}$ $(p, q \geq 0)$ with $\bar\partial f = 0$ in a neighborhood of K_P, one can find $F \in C^\infty_{(p,q)}$ with $\bar\partial F = 0$ in a neighborhood of $K \times D$, so that $f = \mu^* F$ in a neighborhood of K_P.*

Note that K_P is equal to $\mu^{-1}(K \times D)$. The notation μ^* was introduced in section 2.1.

Proof. We first prove (b). Let π be the projection

$$\mathbf{C}^{n+1} \ni (z,w) \to z \in \mathbf{C}^n.$$

Then $\pi \circ \mu$ is the identity. If ω is an open set containing K_P such that $f \in C^\infty_{(p,q)}(\omega)$ and $\bar\partial f = 0$ in ω, then $\pi^* f \in C^\infty_{(p,q)}(\pi^{-1}\omega)$ and $\bar\partial \pi^* f = \pi^* \bar\partial f = 0$ in $\pi^{-1}\omega$, which is a neighborhood of

$$\mu K_P = \{(z,w) \in K \times D, w = P(z)\}.$$

We have $\mu^* \pi^* f = (\pi\mu)^* f = f$ in ω. Let now $\varphi \in C_0^\infty(\pi^{-1}\omega)$ be equal to 1 in a neighborhood of μK_P and set

$$F = \varphi \pi^* f - QG$$

where $Q(z,w) = w - P(z)$ and G is a form of type (p,q) in a neighborhood of $K \times D$ which shall be determined so that $\bar\partial F = 0$. Note that $\mu^* F = (\mu^* \varphi) f = f$ in a neighborhood of K_P. The equation $\bar\partial F = 0$ can be written in the form $Q \bar\partial G = \bar\partial \varphi \wedge \pi^* f$, for $\pi^* f$ is $\bar\partial$-closed; thus we have the equation

$$\bar\partial G = \frac{1}{Q} \bar\partial \varphi \wedge \pi^* f = H.$$

Since $\bar\partial \varphi = 0$ in a neighborhood of μK_P, we have $H \in C^\infty_{(p,q+1)}$ in a neighborhood of $K \times D$, and $\bar\partial H = (1/Q) \bar\partial^2 \varphi \wedge \pi^* f = 0$. Since $K \times D$ is assumed to have the Cousin property, we can find a solution $G \in C^\infty_{(p,q)}$ of the equation $\bar\partial G = H$ in a neighborhood of $K \times D$, and this proves (b).

To prove (a), we take $f \in C^\infty_{(p,q+1)}$ with $\bar\partial f = 0$ in a neighborhood of K_P. According to (b), we can choose $F \in C^\infty_{(p,q+1)}$ so that $\bar\partial F = 0$ in a neighborhood of $K \times D$ and $f = \mu^* F$. Since $K \times D$ has the Cousin property, the equation $\bar{\partial} U = F$ has a solution $\in C^{\infty}_{(p,q)}$ in a neighborhood of $K \times D$. If we set $u = \mu^* U$, it follows that $\bar{\partial} u = \mu^* \bar{\partial} U = \mu^* F = f$ in a neighborhood of K_P. The proof is complete.

Remark. Note that the proof of (a) involves the existence of solutions of the equation $\bar{\partial} G = H$ in a neighborhood of $K \times D$ both for forms H of type $(p, q + 1)$ and $(p, q + 2)$. This is the main reason why we cannot restrict ourselves to forms H of type (0,1).

From Theorem 2.3.3 we know that every closed polydisc has the Cousin property. By repeated application of Lemma 2.7.5 we therefore obtain

Theorem 2.7.6. *Let $\triangle$ be a closed polydisc in $\mathbf{C}^n$, let D be the closed unit disc in $\mathbf{C}$, and let $P_1, \cdots, P_m$ be polynomials. Denote by μ the Oka map*

$$\mathbf{C}^n \ni z \to (z, P_1(z), \cdots, P_m(z)) \in \mathbf{C}^{n+m},$$

and set $K = \{z ; z \in \triangle, \ |P_j(z)| \leq 1, \ j = 1, \cdots, m\} = \mu^{-1}(\triangle \times D^m)$. *Then*

(a) *K has the Cousin property.*

(b) *For every $f \in C^{\infty}_{(p,q)}$ $(p, q \geq 0)$ with $\bar{\partial} f = 0$ in a neighborhood of K, one can find $F \in C^{\infty}_{(p,q)}$ with $\bar{\partial} F = 0$ in a neighborhood of $\triangle \times D^m$ so that $f = \mu^* F$ in a neighborhood of K.*

Proof. A proof by induction over m is obvious in view of Lemma 2.7.5.

We shall now deduce some consequences of Theorem 2.7.6; in particular, we shall complete the proof of Theorem 2.7.3.

Theorem 2.7.7. *Let f be an analytic function in a neighborhood of a compact polynomially convex set K. Then there is a sequence f_j of analytic polynomials such that $f_j \to f$ uniformly on K.*

Proof. By Lemma 2.7.4 we can choose a compact polynomial polyhedron L containing K so that f is analytic in a neighborhood of L. Choose a closed polydisc $\triangle$ with center at 0 so that $\triangle \supset L$, and let

$$L = \{z ; z \in \triangle, |P_j(z)| \leq 1, j = 1, \cdots, m\}.$$

By Theorem 2.7.6 (b) we can find a function F which is analytic in a neighborhood of the polydisc $\triangle \times D^m$ in $\mathbf{C}^{n+m}$ so that

$$f(z) = F(z, P_1(z), \cdots, P_m(z))$$

in a neighborhood of L. But if $F_k(z,w)$ are the partial sums of the power series expansion of $F(z,w)$, we have $F_k \to F$ uniformly in the polydisc $\triangle \times D^m$. Hence

$$F_k(z, P_1(z), \cdots, P_m(z)) \to F(z, P_1(z), \cdots, P_m(z)) = f(z)$$

uniformly on L, which proves the theorem.

End of proof of Theorem 2.7.3. That (ii) implies (i) follows immediately from Theorem 2.7.7. To prove that (iv) $\Rightarrow$ (ii) we set

$$K_1 = \tilde{K} \cap \Omega, \qquad K_2 = \tilde{K} \cap \complement\Omega.$$

Then K_1 is compact by (iv), and K_2 is a closed subset of $\tilde{K}$ and therefore also compact. K_1 and K_2 are disjoint, so we can define a function which is analytic in a neighborhood of $\tilde{K}$ by setting $f = 0$ in a neighborhood of K_1 and $f = 1$ in a neighborhood of K_2. Since $\tilde{K}$ is polynomially convex, it follows from Theorem 2.7.7 that we can find a polynomial g such that $|g - f| < \frac{1}{2}$ on $\tilde{K}$, hence $|g| < \frac{1}{2}$ on $K_1 \supset K$ and $|g| > \frac{1}{2}$ on K_2. Since $K_2 \subset \tilde{K}$, we conclude that K_2 is empty. Using Theorem 2.7.7, we conclude that (i) is valid, hence also (iii) and (ii).

We leave as an exercise to prove that analogues of Corollaries 2.5.6 and 2.5.7 and Theorems 2.5.13 and 2.5.14 are valid for Runge domains. Instead we shall improve part (a) of Theorem 2.7.6.

Theorem 2.7.8. *If Ω is a Runge domain, the equation $\bar{\partial} u = f$ has a solution $u \in C^\infty_{(p,q)}(\Omega)$ for every $f \in C^\infty_{(p,q+1)}(\Omega)$ such that $\bar{\partial} f = 0$ $(p, q \geq 0)$.*

Proof. Let K_j be an increasing sequence of compact subsets of Ω such that $\tilde{K}_j = K_j$ for every j, and every compact subset of Ω is contained in some K_j. According to (a) in Theorem 2.7.6, we can for every j find $u_j \in C^\infty_{(p,q)}(\Omega)$ satisfying the equation $\bar{\partial} u_j = f$ in a neighborhood of K_j. We wish to make the sequence u_j convergent.

First assume that $q > 0$. We claim that the forms u_j can then be chosen so that $u_{j+1} = u_j$ in a neighborhood of K_j for every j. In fact, assume that $u_1, \cdots, u_j$ have already been chosen so that this condition is fulfilled. If $u \in C^\infty_{(p,q)}(\Omega)$ satisfies the equation $\bar{\partial} u = f$ in a neighborhood of K_{j+1}, we have $\bar{\partial}(u_j - u) = f - f = 0$ in a neighborhood of K_j, so by Theorem 2.7.6 again we can choose $v \in C^\infty_{(p,q-1)}(\Omega)$ so that $u_j - u = \bar{\partial} v$ in a neighborhood of K_j. Hence $\bar{\partial}(u + \bar{\partial} v) = \bar{\partial} u = f$ in a neighborhood of K_{j+1} and $u + \bar{\partial} v = u_j$ in a neighborhood of K_j. We can therefore choose $u_{j+1} = u + \bar{\partial} v$. Now it is clear that $u = \lim_{j\to\infty} u_j$ exists, that $u \in C^\infty_{(p,q)}(\Omega)$, and that $\bar{\partial} u = f$.

The case $q = 0$ is slightly more involved. If u is a form of type $(p,0)$,

$$u = \sum_{|I|=p}' u_I \, dz^I,$$

we write $|u| = (\Sigma' |u_I|^2)^{1/2}$. Note that the equation $\bar\partial u = 0$ means that $\bar\partial u_I = 0$ for every I, that is, all u_I are analytic. If we argue as above, using Theorem 2.7.7 instead of appealing to Theorem 2.7.6 a second time, it follows that the forms $u_j \in C^\infty_{(p,0)}(\Omega)$ can be chosen so that $\bar\partial u_j = f$ in a neighborhood of K_j and $|u_{j+1} - u_j| < 2^{-j}$ in K_j for every j. Since the coefficients of $u_{j+1} - u_j$ are analytic in K_j, we conclude that $u = \lim_{j\to\infty} u_j$ exists and that $u - u_j$ has analytic coefficients in the interior of K_j for every j. Hence $u \in C^\infty_{(p,0)}(\Omega)$ and $\bar\partial u = f$ in Ω.

In the case of one complex variable we found that a domain is a Runge domain precisely if it is simply connected. A topological characterization of Runge domains in $\mathbf{C}^n$ is not possible when $n > 1$. In fact, at the end of section 2.4 we have given an example of a domain which is homeomorphic to a ball and is not even a domain of holomorphy. However, we shall now prove that there are topological restrictions on Runge domains.

Lemma 2.7.9. *Let Ω be an open set in $\mathbf{C}^n$ where the equation $\bar\partial u = f$ has a solution $u \in C^\infty_{(p,q)}(\Omega)$ for all $f \in C^\infty_{(p,q+1)}(\Omega)$ with $\bar\partial f = 0$ $(p, q \ge 0)$. If f is a differential form of degree $r > 0$ with coefficients in $C^\infty(\Omega)$ and if $df \in C^\infty_{(r+1,0)}$, one can then find a form $f' \in C^\infty_{(r,0)}(\Omega)$ and a form g of degree $r - 1$ with coefficients in $C^\infty(\Omega)$ so that $f - f' = dg$; note that this implies that $df' = df$, hence that $\bar\partial f' = 0$.*

Proof. We can write f in the form

$$f = \sum_0^r f_{r-q,q},$$

where $f_{r-q,q} \in C^\infty_{(r-q,q)}(\Omega)$. By induction over k we shall prove that the theorem is true if $f_{r-q,q} = 0$ when $q > k$. This statement is obvious if $k = 0$ and coincides with the theorem if $k = r$. Assume that $0 < k \le r$ and that the statement has already been proved when k is replaced by $k - 1$. Since $\bar\partial f_{r-k,k}$ is the only term of type $(r - k, k + 1)$ in df, we have $\bar\partial f_{r-k,k} = 0$, so by hypothesis there is a form $g' \in C^\infty_{(r-k,k-1)}(\Omega)$ such that $\bar\partial g' = f_{r-k,k}$. Then

$$f - dg' = \sum_0^{k-1} f_{r-q,q} - \partial g'$$

and, since $\partial g' \in C^\infty_{(r-k+1,k-1)}(\Omega)$, it follows from the inductive hypothesis that $f - dg' = f' + dg''$, where $f' \in C^\infty_{(r,0)}(\Omega)$ and g'' is of degree $r - 1$. This completes the proof.

Now recall that by the de Rham theorem (see also section 7.5) there is a natural isomorphism

$H^r(\Omega,\mathbf{C}) \approx \{C^\infty$ forms f of degree r with $df = 0\}/\{dg$ where g is a C^∞ form of degree $r - 1\}$;

here $H^r(\Omega,\mathbf{C})$ is the rth cohomology group of Ω with complex coefficients. When the hypotheses of Lemma 2.7.9 are fulfilled, there is a form $f \in C^\infty_{(r,0)}(\Omega)$, with $df = 0$, in every residue class. Further, if such a form can be written $f = dg$ with g of degree $r - 1$, then g can be chosen in $C^\infty_{(r-1,0)}(\Omega)$, and $\bar\partial g = 0$. Let us call a form $h \in C^\infty_{(s,0)}(\Omega)$ with $\bar\partial f = 0$ a *holomorphic form* of degree s. (Note that all coefficients of a holomorphic form are holomorphic.) Then the remarks just made prove

Theorem 2.7.10. *Let Ω be an open set in $\mathbf{C}^n$ where the equation $\bar\partial u = f$ has a solution $u \in C^\infty_{(p,q)}(\Omega)$ for all $f \in C^\infty_{(p,q+1)}(\Omega)$ with $\bar\partial f = 0$ $(p, q \geq 0)$. Then*

$H^r(\Omega,\mathbf{C}) \approx \{$holomorphic forms f of degree r with $\partial f = 0\}/\{\partial g$ where g is a holomorphic form of degree $r - 1\}$.

Hence $H^r(\Omega,\mathbf{C}) = 0$ when $r > n$.

Theorem 2.7.10 applies in particular to Runge domains. We shall then prove that also the nth cohomology group is 0:

Theorem 2.7.11. *If Ω is a Runge domain in $\mathbf{C}^n$, then $H^r(\Omega,\mathbf{C}) = 0$ when $r \geq n$.*

Proof. We have to prove that every holomorphic form of degree n is the exterior differential of a holomorphic form of degree $n - 1$. Now a holomorphic form of degree n can be written

$$f\, dz_1 \wedge \cdots \wedge dz_n \tag{2.7.1}$$

where $f \in A(\Omega)$, and a holomorphic form of degree $n - 1$ can be written

$$\sum_1^n (-1)^{j-1} f_j\, dz_1 \wedge \cdots \wedge d\hat{z}_j \wedge \cdots \wedge dz_n \tag{2.7.2}$$

where $d\hat{z}_j$ means that dz_j shall be omitted, and $f_j \in A(\Omega)$. The differential of the form (2.7.2) is equal to the form (2.7.1) if

$$f = \sum_1^n \partial f_j/\partial z_j.$$

When f is analytic in the whole space, we can simply choose $f_2 = \cdots = f_n = 0$ and $f_1(z) = \int_0^{z_1} f(t, z_2, \cdots, z_n)\, dt$ (complex contour integration), and entire functions are dense in $A(\Omega)$ since Ω is a Runge domain. Hence differentials of $(n - 1)$-forms are dense in the set of all closed n-forms, and we shall prove in section 7.5 that this implies that $H^n(\Omega,\mathbf{C}) = 0$.

Since a polynomially convex set in $\mathbf{C}^n$ has a fundamental set of neighborhoods which are Runge domains, Theorem 2.7.11 and an elementary fact concerning Čech cohomology which we shall prove in section 7.5 give the following result:

Theorem 2.7.12. *If K is a compact polynomially convex set in $\mathbf{C}^n$, then $H^r(K,\mathbf{C}) = 0$ when $r \geq n$.*

We shall finally give an example which shows that $H^r(\Omega,\mathbf{C})$ does not vanish for all Runge domains Ω when $r < n$. Set

$$\Omega = \{z\,; z \in \mathbf{C}^n, |z_1 \cdots z_n - 1| < 1, |z_j| < 2, j = 1, \cdots, n\}.$$

This is a bounded Runge domain (compare the proof of Theorem 2.5.13). Let $1 \leq r < n$ and consider the form

$$f = (z_1 \cdots z_r)^{-1}\, dz_1 \wedge \cdots \wedge dz_r$$

which is a closed holomorphic r-form in Ω, since $z_1 \cdots z_n \neq 0$ in Ω. Let γ be the cycle in Ω defined by

$$z_j = e^{i\theta_j}, \quad j = 1, \cdots, r; \qquad z_{r+1} = e^{-i(\theta_1 + \cdots + \theta_r)}, \quad z_j = 1, j > r + 1.$$

Here the parameters θ_j vary from 0 to 2π. Now we have

$$\int_\gamma f = \int i^r\, d\theta_1 \cdots d\theta_r = (2\pi i)^r \neq 0.$$

Hence f is not the differential of any form of degree $r - 1$, which proves that $H^r(\Omega,\mathbf{C}) \neq 0$.

Notes. We refer readers who are not familiar with differential forms to de Rham [1] for the basic definitions assumed in section 2.1.—Sections 2.2 and 2.4 contain only classical results, which mostly can be found in Osgood [1], but the presentation owes much to the introductory chapter of Malgrange [1].—Theorem 2.3.2 is a famous theorem of Hartogs. The refined version Theorem 2.3.2′ is due to Bochner [1]. The usual proofs of these results depend on deformations of a curve to which the Cauchy integral formula is applied. As pointed out by Ehrenpreis [1], what is involved is really a theorem on existence of compactly supported solutions to the Cauchy–Riemann equations, and so we start from such a result instead. (Similar questions have been discussed for complex manifolds by Kohn and Rossi [1].) Theorem 2.3.3, the Poincaré lemma for the $\bar\partial$ operator, is often referred to as the Dolbeault–Grothendieck lemma.—The characterization of domains of holomorphy in Theorem 2.5.5 is due to Cartan and Thullen [1]. Theorem 2.5.10 is due to Bochner (see Bochner–Martin [1] for another proof). A simpler but less elementary proof can be based on the study of envelopes of holomorphy in section 5.4.—That domains of holomorphy enjoy convexity properties was already found by Levi and Hartogs. The presentation in section 2.6 follows Bremermann [1] and Lelong [1]. Theorem 2.6.13 is essentially due to Lewy [1].—The methods and results of section 2.7 are basically due to Oka [1], although his paper looks rather different formally since he studies the Cousin problems and not the Cauchy–Riemann equations. The consideration of differential forms of higher order in Lemma 2.7.5 (or equivalently the study of cohomology groups of higher order) which is now standard is also an essential simplification.—Theorem 2.7.10 is proved in Cartan [1] for Stein manifolds. We shall do so in section 5.2. Theorem 2.7.11 is due to Serre [2], and Theorem 2.7.12 was published by Browder [1]. The example at the end of section 2.7 has been taken from Behnke and Stein [1].

Chapter III

APPLICATIONS TO COMMUTATIVE BANACH ALGEBRAS

Summary. We first recall the basic notions in the theory of Banach algebras. Inasmuch as the presentation is short, a reader who is totally unfamiliar with the subject may need to consult a more detailed text, for example Naimark [1] or Loomis [1]. In section 3.2 we then prove that analytic functions of several complex variables operate on the space of Gelfand transforms. We also show that the Shilov boundary is determined by local conditions.

3.1. Preliminaries. We shall first recall some basic facts and definitions concerning commutative Banach algebras.

Definition 3.1.1. *An algebra B (over the complex numbers) is called a Banach algebra if there is a norm given in B under which B is a Banach space and*

$$\|fg\| \leq \|f\| \, \|g\|, \qquad f, g \in B.$$

We shall only consider *commutative Banach algebras with unit element*, denoted by e. This assumption will therefore not be stated explicitly every time.

One of the main objectives of the theory is to study to what extent it is possible to represent the algebra by an algebra of continuous functions on a compact space. Let K be a compact space and $C(K)$ the algebra of continuous complex valued functions on K. Suppose that

$$B \ni f \to Tf \in C(K)$$

is a continuous representation of B, that is, that T commutes with the algebraic operations of the algebra and

$$\sup_K |Tf| \leq C\|f\|$$

for some constant C. Since $T(f^n) = (Tf)^n$, it follows that

$$\sup_K |Tf| \le C^{1/n} \|f^n\|^{1/n} \le C^{1/n} \|f\|.$$

Hence

$$\sup |Tf| \le \varliminf_{n\to\infty} \|f^n\|^{1/n} \le \|f\|. \tag{3.1.1}$$

We shall now prove that among the mappings of the type just considered there is one from which all others can be obtained.

Definition 3.1.2. *A linear form m on B is called a multiplicative linear functional if it is continuous, not identically* 0, *and*

$$m(fg) = m(f)m(g) \qquad (f, g \in B).$$

We denote by M_B or simply M the set of all multiplicative linear functionals on B with the weakest topology which makes the mapping

$$M \ni m \to m(f) \in \mathbf{C}$$

continuous for every $f \in B$.

The definition of the topology means that a fundamental system of neighborhoods of $m_0 \in M$ can be obtained by taking finite intersections of neighborhoods of the form $\{m; |m(f) - m_0(f)| < \varepsilon\}$, where $f \in B$ and $\varepsilon > 0$. It is clear that the condition $m \neq 0$ is equivalent to $m(e) = 1$.

Theorem 3.1.3. *M_B is a compact Hausdorff space.*

Proof. Application of (3.1.1) to the mapping $f \to m(f)$ for a fixed $m \in M$ shows that $|m(f)| \le \|f\|$ for every $f \in B$. Let D_f be the disc $\{z; z \in \mathbf{C}, |z| \le \|f\|\}$. Then the map

$$M \ni m \to \{m(f)\}_{f\in B} \in \prod_{f\in B} D_f = D$$

is a homeomorphism of M onto the set of all $z = \{z_f\} \in D$ such that $z_e = 1$ and

$$z_{af+bg} = az_f + bz_g, \quad z_{fg} = z_f z_g, \qquad f, g \in B, a, b \in \mathbf{C},$$

which is a closed subset of D and therefore compact. This proves the theorem.

Definition 3.1.4. *The continuous function $\hat{f}$ on M defined by*

$$\hat{f}(m) = m(f), \qquad m \in M$$

is called the Gelfand transform of $f \in B$. The mapping $B \ni f \to \hat{f} \in C(M)$ is called the Gelfand representation of B.

Now consider again an arbitrary continuous representation

$$B \ni f \to Tf \in C(K),$$

where K is a compact space. Since $(Te)^2 = T(e^2) = Te$, the continuous function Te on K only takes the two values 0 and 1. Put $K_0 = \{k; k \in K, (Te)(k) = 0\}$ and $K_1 = \{k; k \in K, (Te)(k) = 1\}$. Then K_0 and K_1 are compact and disjoint, and $Tf = 0$ in K_0 for every $f \in B$. Only the restriction of Tf to K_1 is therefore of any interest. For every $k \in K_1$, the map $B \ni f \to (Tf)(k)$ defines an element $m \in M$, which we denote by $\varphi(k)$. The definition of the topology in M and the fact that Tf is continuous for every $f \in B$ imply that φ is continuous, and since $Tf = \hat{f} \circ \varphi$ on K_1 we have found a complete description of all representations of B by continuous functions in terms of the Gelfand representation.

We next turn to theorems on existence of multiplicative linear functionals.

Definition 3.1.5. *If $f \in B$, the spectrum $\sigma(f)$ of f is defined as the set of all $\lambda \in \mathbf{C}$ such that $f - \lambda e$ has no inverse.*

Theorem 3.1.6. *For every $f \in B$ we have*

$$\sigma(f) = \{\hat{f}(m); m \in M\}, \quad \sup_{m \in M} |\hat{f}(m)| = \lim_{n \to \infty} \|f^n\|^{1/n}. \tag{3.1.2}$$

The proof of this theorem will require some preparations. Now we only note that $\{\hat{f}(m); m \in M\} \subset \sigma(f)$. In fact, if $\lambda \notin \sigma(f)$, we can choose $g \in B$ so that $g(f - \lambda e) = e$, hence $\hat{g}(\hat{f} - \lambda) = 1$, which proves that $\hat{f} \neq \lambda$.

Lemma 3.1.7. *If g^{-1} exists, then $(g - \lambda h)^{-1}$ exists when $|\lambda| \, \|g^{-1}h\| < 1$, and is a continuous function of λ there. If ω is a relatively compact open subset of this disc, bounded by a finite number of C^1 arcs, then*

$$\int_{\partial\omega} (g - \lambda h)^{-1} \varphi(\lambda) \, d\lambda = 0$$

if φ is analytic in ω and belongs to $C^1(\bar{\omega})$.

Proof. If $g^{-1}h = H$, then

$$I(\lambda) = g^{-1} \sum_0^\infty \lambda^n H^n$$

converges normally in the disc in question, and $I(\lambda)(g - \lambda h) =$

$I(\lambda)g(e - \lambda H) = e$. This proves the lemma since we can integrate term by term over $\partial\omega$ after multiplying by $\varphi(\lambda)$.

Lemma 3.1.8. *If* $(e - \lambda f)^{-1}$ *exists when* $|\lambda| \leq R$, *then*

$$R^n\|f^n\| \leq \sup_{|\lambda|=R} \|(e - \lambda f)^{-1}\|, \qquad n \geq 0. \tag{3.1.3}$$

Proof. From Lemma 3.1.7 it follows that

$$(2\pi i)^{-1} \int_{|\lambda|=r} (e - \lambda f)^{-1} \lambda^{-n-1} \, d\lambda$$

is independent of r when $0 < r \leq R$, and when $r\|f\| < 1$ we find that the integral is equal to f^n by integrating the series expansion. This proves the lemma.

We can now prove another part of Theorem 3.1.6. Let $1/R > \sup_{z\in\sigma(f)} |z|$. Then $(e - \lambda f)^{-1}$ exists when $|\lambda| \leq R$, so it follows from (3.1.3) that

$$R \varlimsup_{n\to\infty} \|f^n\|^{1/n} \leq 1.$$

Hence

$$\varlimsup_{n\to\infty} \|f^n\|^{1/n} \leq \sup_{z\in\sigma(f)} |z|. \tag{3.1.4}$$

If we can prove that $\sigma(f) \subset \{\hat{f}(m); m \in M\}$, the proof of Theorem 3.1.6 will thus be complete, for (3.1.1) and (3.1.4) then give

$$\sup_{m\in M} |\hat{f}(m)| \leq \varliminf_{n\to\infty} \|f^n\|^{1/n} \leq \varlimsup_{n\to\infty} \|f^n\|^{1/n} \leq \sup_{z\in\sigma(f)} |z| = \sup_{m\in M} |\hat{f}(m)|,$$

so that equality must hold all the way.

Lemma 3.1.9. *No* $f \in B$ *has an empty spectrum.*

Proof. If f has an empty spectrum, then $(e - \lambda f)^{-1}$ exists for every $\lambda \in \mathbf{C}$ and $\|(e - \lambda f)^{-1}\| \leq |\lambda|^{-1}\|f^{-1}\| \, \|(e - \lambda^{-1}f^{-1})^{-1}\| = O(|\lambda|^{-1})$ when $\lambda \to \infty$. This contradicts (3.1.3) when $n = 0$.

Lemma 3.1.10. *If a Banach algebra B is a field, then B is isomorphic to the field of complex numbers.*

Proof. By Lemma 3.1.9 we can for every $f \in B$ find $\lambda \in \mathbf{C}$ so that $f - \lambda e$ has no inverse. But if B is a field, this means that $f - \lambda e = 0$. Hence B consists of all complex multiples of e.

Recall now that a vector space $I \subset B$ is called an *ideal* if $BI \subset I$, that is, if $gf \in I$ for all $g \in B$ and $f \in I$. If $I \neq B$, we say that the ideal is proper. No element $f \in B$ with $\|e - f\| < 1$ can belong to a proper ideal I, for $f = e - (e - f)$ would then have an inverse $g \in B$, hence $e = gf \in I$ so that $I = B$. The closure $\bar{I}$ of a proper ideal is therefore a proper ideal.

Lemma 3.1.11. *If I is a maximal ideal in B, that is, I is a proper ideal contained in no other proper ideal in B, then I is closed and B/I is isomorphic to the complex number field.*

Proof. Since $\bar{I}$ is a proper ideal containing I, it is clear that I is closed. The quotient algebra B/I is therefore a Banach algebra with the quotient norm, and it contains no ideals $\neq \{0\}$ and $\neq B/I$ since the inverse image in B of an ideal in B/I is an ideal in B containing I. But this means that B/I is a field, because the ideal generated by an arbitrary non-zero element in B/I contains the identity in B/I. Hence the lemma follows from Lemma 3.1.10.

If I is a maximal ideal, the mapping $B \ni f \to B/I$ can therefore be regarded as a multiplicative linear functional m, and $I = \{f \in B; m(f) = 0\}$. Conversely, if $m \in M$, it is clear that $I = \{f \in B; m(f) = 0\}$ is a proper ideal in M, and since B/I is of dimension 1, the ideal is maximal. Hence there is a one-to-one correspondence between maximal ideals in B and multiplicative linear functionals on B. The space M is therefore often called the maximal ideal space of B.

Theorem 3.1.12. *If I is a proper ideal in B, there exists an element $m \in M$ such that $m(f) = 0$ for every $f \in I$.*

Proof. Since no proper ideal contains e, an application of Zorn's lemma (see Loomis [1], p. 2) shows that every proper ideal is contained in a maximal ideal. But this ideal is the set of zeros of a multiplicative linear functional.

End of proof of Theorem 3.1.6. If $\lambda \in \sigma(f)$, then the ideal generated by $(f - \lambda e)$ is different from B, so by Theorem 3.1.12 one can find $m \in M$ so that m vanishes on this ideal. In particular, $m(f - \lambda e) = 0$, that is, $\lambda = m(f)$. Hence $\sigma(f) \subset \{\hat{f}(m); m \in M\}$, and by the remarks following Lemma 3.1.8, this completes the proof.

Instead of considering the spectrum of a single element in B, one can study the spectrum of a set of elements:

Definition 3.1.13. *If $f_1, \cdots, f_n \in B$, the joint spectrum $\sigma(f_1, \cdots, f_n)$ of these elements is defined as the set of all $\lambda \in \mathbf{C}^n$ such that the ideal generated by $(f_1 - \lambda_1 e), \cdots, (f_n - \lambda_n e)$ is different from B.*

Theorem 3.1.6 has an immediate extension:

Theorem 3.1.14. *For arbitrary* $f_1, \cdots, f_n \in B$ *we have*

$$\sigma(f_1, \cdots, f_n) = \{(\hat{f}_1(m), \cdots, \hat{f}_n(m)), m \in M\}.$$

Proof. If $\lambda \in \sigma(f_1, \cdots, f_n)$, it follows from Theorem 3.1.12 that one can find $m \in M$ such that $m(f_j - \lambda_j e) = 0$ for $j = 1, \cdots, n$, that is, $m(f_j) = \lambda_j$. On the other hand, if $\lambda \notin \sigma(f_1, \cdots, f_n)$ we can choose $g_j \in B$ so that

$$\sum_1^n g_j(f_j - \lambda_j e) = e.$$

Hence $\Sigma_1^n \hat{g}_j(\hat{f}_j - \lambda_j) = 1$, which proves that $(\hat{f}_1(m), \cdots, \hat{f}_n(m)) \neq (\lambda_1, \cdots, \lambda_n)$ for every $m \in M$.

We shall say that B is generated by the elements $f_1, \cdots, f_n \in B$ if the smallest closed subalgebra of B containing $f_1, \cdots, f_n$ is equal to B. In other words, the elements $P(f_1, \cdots, f_n)$ where P is a polynomial with complex coefficients shall be dense in B. (In defining $P(f_1, \cdots, f_n) = \Sigma\, a_\alpha f^\alpha$, we set $f^0 = e$.) The maximal ideal space of a finitely generated algebra is particularly simple to describe.

Theorem 3.1.15. *Let* B *be generated by* $f_1, \cdots, f_n$. *Then the map*

$$\varphi : M \ni m \to (\hat{f}_1(m), \cdots, \hat{f}_n(m)) \in \mathbf{C}^n$$

is a homeomorphism of M *on a compact polynomially convex set* K *in* $\mathbf{C}^n$, *which is by Theorem* 3.1.14 *the joint spectrum of the system of generators* $f_1, \cdots, f_n$. *If* $f \in B$, *then* $\hat{f} \circ \varphi^{-1}$ *can be uniformly approximated by polynomials on* K.

Note that the joint spectrum of a system of elements which are not generators of B need not be polynomially convex.

Proof. The map φ is continuous by definition of the topology in M, and φ is $1 - 1$ because, if $m \in M$,

$$m(P(f_1, \cdots, f_n)) = P(\hat{f}_1(m), \cdots, \hat{f}_n(m)),$$

so that $(\hat{f}_1(m), \cdots, \hat{f}_n(m))$ determines m uniquely on a dense set in B and therefore on all of B. Since M is compact, it follows that K is compact and that φ is a homeomorphism. Now let $z \in \tilde{K}$ (this notation was introduced before Definition 2.7.2). To prove that $z \in K$ we must show that the map

$$f_j \to z_j$$

can be extended to a multiplicative linear form on B. Clearly such a form must for every polynomial P map $P(f_1, \cdots, f_n)$ on $P(z_1, \cdots, z_n)$, so what we have to prove is the continuity of this map, that is, that

$$|P(z_1, \cdots, z_n)| \leq \|P(f_1, \cdots, f_n)\|$$

for all polynomials P. Now we have

$$\begin{aligned} |P(z_1, \cdots, z_n)| &\leq \sup_{\zeta \in K} |P(\zeta_1, \cdots, \zeta_n)| = \sup_{m \in M} |m(P(f_1, \cdots, f_n))| \\ &\leq \|P(f_1, \cdots, f_n)\|. \end{aligned}$$

This proves the theorem, for the final statement follows from the fact that polynomials in $f_1, \cdots, f_n$ are dense in B and that $\sup_M |\hat{g}| \leq \|g\|$ for every $g \in B$.

Combination of Theorem 3.1.15 and Theorem 2.7.12 gives

Corollary 3.1.16. *If B has n generators, then $H^k(M,\mathbf{C}) = 0$ when $k \geq n$.*

Example. The continuous functions on the n-sphere have $n+1$ generators by Weierstrass' approximation theorem but not n generators in view of the corollary, for M is the n-sphere itself.

To give an example of Theorem 3.1.15, we choose a compact set $K \subset \mathbf{C}^n$ and let B be the closure $C_A(K)$ in $C(K)$ of the restrictions to K of analytic polynomials in $\mathbf{C}^n$. ($C(K)$ is the space of continuous functions on K with the maximum norm.) Then B is a Banach algebra with n generators $z_1, \cdots, z_n$. The maximal ideal space can be identified with $\tilde{K}$, and the Gelfand representation consists in extending the functions in $C_A(K)$ to functions in $C_A(\tilde{K})$. Thus these spaces are isomorphic.

The maximum ideal space can be regarded as the largest space on which the algebra A can be represented, but as shown by the previous example it may happen that the Gelfand transform of all elements in the algebra are determined by their restrictions to a certain subset of M.

Definition 3.1.17. *A closed subset M_0 of M is called a boundary for M (with respect to B) if*

$$\sup_{M_0} |\hat{f}| = \sup_{M} |\hat{f}|, \qquad f \in B.$$

Theorem 3.1.18. *There is a boundary S for M which is contained in every boundary. One calls S the Shilov boundary for M (with respect to B).*

The intersection of all boundaries consists of all $m_0 \in M$ such that $M \setminus V$ is not a boundary for any neighborhood V of m_0, that is, such that for every V one can find $f \in B$ so that $\sup_{M \setminus V} |\hat{f}| < \sup_V |\hat{f}|$. To prove that the set of all such m_0 is indeed a boundary, we need a lemma.

Lemma 3.1.19. *Let $f_1, \cdots, f_n \in B$ and set*

$$U = \{m \in M; |\hat{f}_i(m)| < 1, i = 1, \cdots, n\}.$$

Then either U intersects every boundary of M or else $M_0 \setminus U$ is a boundary of M for every boundary M_0 of M.

Proof. Assume that M_0 but not $M_0 \setminus U$ is a boundary for M. Then we can choose $f \in B$ so that $\sup_M |\hat{f}| = 1$ but $\sup_{M_0 \setminus U} |\hat{f}| < 1$. Replacing f by a power of f, we may assume that $|\hat{f}| < \varepsilon$ in $M_0 \setminus U$, where $\varepsilon > 0$ is chosen so small that $\varepsilon \sup_M |\hat{f}_i| < 1$, $i = 1, \cdots, n$. Then we have $|\hat{f}\hat{f}_i| < 1$ in U by the definition of U and in $M_0 \setminus U$ by the choice of ε. Hence $|\hat{f}\hat{f}_i| < 1$ in M_0; and, since M_0 is a boundary, this implies that $|\hat{f}\hat{f}_i| < 1$ on M, $i = 1, \cdots, n$. Every point where $|\hat{f}| = 1$ must therefore belong to U, so U must intersect every boundary of M.

Proof of Theorem 3.1.18. Let S be the intersection of all boundaries, that is, the set of points described immediately after the statement of the theorem. Let $f \in B$ and assume that $|\hat{f}| < 1$ on S. We have to prove that $|\hat{f}| < 1$ on M. To do so, we form the compact set $M' = \{m \in M; |\hat{f}(m)| \geq 1\}$. For every $m \in M'$, we can by the definition of S find a boundary M_0 so that $m \notin M_0$, and therefore a neighborhood U of m of the type discussed in the lemma so that $U \cap M_0 = \emptyset$. By the Borel–Lebesgue lemma we can cover M' by a finite number of such sets $U_1, \cdots, U_r$. Since M is a boundary, it follows by repeated use of the lemma that $M \setminus (\cup_1^r U_j)$ is a boundary. But since $|\hat{f}| < 1$ there, we obtain $|\hat{f}| < 1$ on M, which completes the proof.

Example. The Shilov boundary of a polydisc is its distinguished boundary.

3.2 Analytic functions of elements in a Banach algebra. The following extension of a part of statement (b) in Theorem 2.7.6 is the central result of the section.

Theorem 3.2.1. *Let $f_1, \cdots, f_n \in B$ and let φ be an analytic function in a neighborhood of $\sigma(f_1, \cdots, f_n)$ in $\mathbf{C}^n$. Then one can find a finite number of elements $f_{n+1}, \cdots, f_N \in B$ and a function Φ which is analytic in a neighborhood of the polydisc*

$$\{z\,;z\in \mathbf{C}^N, |z_j| \leq \|f_j\|, j = 1,\cdots,N\}$$

such that $\varphi(\hat{f}_1,\cdots,\hat{f}_n) = \Phi(\hat{f}_1,\cdots,\hat{f}_N)$.

Before the proof we give the main application

Theorem 3.2.2. *Let* $f_1,\cdots,f_n \in B$ *and let* φ *be an analytic function in a neighborhood of* $\sigma(f_1,\cdots,f_n)\in \mathbf{C}^n$. *Then one can find* $g \in B$ *such that*

$$\hat{g} = \varphi(\hat{f}_1,\cdots,\hat{f}_n).$$

In other words, analytic functions operate on the space of Gelfand transforms.

Proof. Choose $f_{n+1},\cdots,f_N$ and Φ according to Theorem 3.2.1. We have

$$\Phi(z) = \sum_\alpha a_\alpha z^\alpha, \quad \sum_\alpha |a_\alpha| R^\alpha < \infty \qquad (R_j = \|f_j\|),$$

where $z = (z_1,\cdots,z_N)$. Hence

$$g = \sum_\alpha a_\alpha f^\alpha$$

exists with norm convergence, and $\hat{g} = \Sigma_\alpha a_\alpha \hat{f}^\alpha = \Phi(\hat{f}_1,\cdots,\hat{f}_N) = \varphi(\hat{f}_1,\cdots,\hat{f}_n)$.

The proof of Theorem 3.2.1 will result from two lemmas.

Lemma 3.2.3. *Let* Ω *be an open set in* $\mathbf{C}^n$ *which contains* $\sigma_B(f_1,\cdots,f_n)$. *Then there exists a finitely generated closed subalgebra* B' *of* B *such that* $f_1,\cdots,f_n \in B'$ *and* $\sigma_{B'}(f_1,\cdots,f_n) \subset \Omega$.

Proof. First note that it is trivial that $\sigma_{B'}(f_1,\cdots,f_n)$ decreases when B' increases. It is therefore sufficient to prove that for every $z \notin \sigma_B(f_1,\cdots,f_n)$ one can choose B' such that $z \notin \sigma_{B'}(f_1,\cdots,f_n)$. To do so, we note that the hypothesis means that $e = \Sigma_1^n g_j(f_j - z_j e)$ for suitable $g_j \in B$. If B' is the closed algebra generated by $f_1,\cdots,f_n, g_1,\cdots,g_n$, it follows that $z \notin \sigma_{B'}(f_1,\cdots,f_n)$.

Choose $f_{n+1},\cdots,f_\nu$ so that $f_1,\cdots,f_\nu$ generate the algebra B' in Lemma 3.2.3, and let π be the projection $(z_1,\cdots,z_\nu) \to (z_1,\cdots,z_n)$ of $\mathbf{C}^\nu$ on $\mathbf{C}^n$.

Lemma 3.2.4. *There exist polynomials* $P_k, k = 1,\cdots,\mu$, *in* ν *variables such that*

$$z \in \mathbf{C}^\nu, |z_j| \leq \|f_j\|, j = 1,\cdots,\nu, |P_k(z)| \leq \|P_k(f_1,\cdots,f_\nu)\|,$$
$$k = 1,\cdots,\mu \Rightarrow \pi z \in \Omega.$$

Proof. If $\pi z \notin \sigma_{B'}(f_1, \cdots, f_n)$, the map $f_j \to z_j, j = 1, \cdots, v$ cannot be extended to a multiplicative linear functional on B'. Hence there is a polynomial P in v variables such that

$$|P(z_1, \cdots, z_v)| > \|P(f_1, \cdots, f_v)\|.$$

This inequality must also hold in a neighborhood of z. Since the set $\{z; |z_j| \leq \|f_j\|, j = 1, \cdots, v\}$ is compact, an application of the Borel–Lebesgue lemma shows that polynomials $P_1, \cdots, P_\mu$ with the required properties can be found.

Proof of Theorem 3.2.1. Let Ω be a neighborhood of $\sigma_B(f_1, \cdots, f_n)$ where φ is analytic, and choose $f_{n+1}, \cdots, f_v$ and $P_1, \cdots, P_\mu$ according to the previous lemmas. Set $f_{v+k} = P_k(f_1, \cdots, f_v), k = 1, \cdots, \mu$ and set $N = v + \mu$. By Lemma 3.2.4 the function $z \to \varphi(\pi z)$ is analytic in a neighborhood of

$$K = \{z; z \in \mathbf{C}^v, |z_j| \leq \|f_j\|, j = 1, \cdots, v, |P_k(z)| \leq \|f_{k+v}\|, k = 1, \cdots, \mu\}.$$

Hence it follows from (b) in Theorem 2.7.6 that we can find a function Φ which is analytic in a neighborhood of the polydisc

$$\{z; z \in \mathbf{C}^N, |z_j| \leq \|f_j\|, j = 1, \cdots, N\}$$

such that

$$\Phi(z_1, \cdots, z_v, P_1(z), \cdots, P_\mu(z)) = \varphi(\pi z)$$

for all z in a neighborhood of K. In particular, if $m \in M$ we can choose $z_j = \hat{f}_j(m) = m(f_j), j = 1, \cdots, v$, which gives $P_j(z) = m(f_{j+v}) = \hat{f}_{j+v}(m)$. Hence $z \in K$ and we obtain

$$\Phi(\hat{f}_1, \cdots, \hat{f}_N) = \varphi(\hat{f}_1, \cdots, \hat{f}_n),$$

which proves Theorem 3.2.1.

As an application of Theorem 3.2.2 we shall prove an "implicit function theorem" for a Banach algebra. This concerns the solution of an algebraic equation

$$\sum_0^n a_k w^k = 0, \tag{3.2.1}$$

with coefficients $a_k \in B$ and the unknown $w \in B$. This equation implies the corresponding equation for the Gelfand transforms,

$$\sum_0^n \hat{a}_k \hat{w}^k = 0, \tag{3.2.2}$$

which for every $m \in M$ is an algebraic equation with complex coefficients for $\hat{w}(m)$. A necessary condition for the existence of a solution to (3.2.1) is therefore that (3.2.2) has a solution which depends continuously on m. Conversely, we shall prove

Theorem 3.2.5. *Suppose that there is a continuous function h on M such that*

$$(3.2.3) \qquad \sum_0^n \hat{a}_k h^k = 0, \qquad \sum_1^n k\hat{a}_k h^{k-1} \neq 0 \quad \text{at every point of } M.$$

Then there is a solution w of (3.2.1) with $\hat{w} = h$.

The proof will proceed in two steps. First we shall only determine w so that (3.2.2) holds, and afterward we correct w so that (3.2.1) is valid.

Lemma 3.2.6. *One can find a finite number of elements $a_{n+1}, \cdots, a_\nu \in B$ such that $h(m_1) = h(m_2)$ for all $(m_1, m_2) \in M \times M$ with $\hat{a}_j(m_1) = \hat{a}_j(m_2)$, $j = 0, \cdots, \nu$.*

Proof. For every point $(m_1{}^0, m_2{}^0) \in M \times M$ with $m_1{}^0 \neq m_2{}^0$ we can choose $a \in B$ so that $\hat{a}(m_1{}^0) \neq \hat{a}(m_2{}^0)$, hence $\hat{a}(m_1) \neq \hat{a}(m_2)$ in a neighborhood of $(m_1{}^0, m_2{}^0)$. On the other hand, consider a diagonal element (m^0, m^0). Choose $\delta > 0$ so that the polynomial $\Sigma_0^n \hat{a}_k(m^0)z^k$ only has the simple zero $z = h(m^0)$ in the disc $|z - h(m^0)| \leq \delta$. Let V be a neighborhood of m^0 such that $|h(m) - h(m^0)| \leq \delta$ and the equation $\Sigma_0^n \hat{a}_k(m)z^k = 0$ has only one root with $|z - h(m^0)| \leq \delta$ when $m \in V$. If $m \in V$, it follows that $h(m)$ is the only zero of $\Sigma_0^k \hat{a}_k(m)z^k$ with $|z - h(m^0)| \leq \delta$, so that, if $(m_1, m_2) \in V \times V$ and $\hat{a}_k(m_1) = \hat{a}_k(m_2)$, $k = 0, \cdots, n$, it follows that $h(m_1) = h(m_2)$. Since $M \times M$ is compact, we can now use the Borel–Lebesgue lemma to find a finite number of elements $a_{n+1}, \cdots, a_\nu$ with the required properties.

Lemma 3.2.7. *Let $a_{n+1}, \cdots, a_\nu$ be chosen according to Lemma 3.2.6. Then there is a function H which is analytic in a neighborhood of $\sigma(a_0, \cdots, a_\nu)$ in $\mathbf{C}^{\nu+1}$ such that $h = H(\hat{a}_0, \cdots, \hat{a}_\nu)$.*

Proof. According to Lemma 3.2.6 there is a uniquely defined function H on $\sigma(a_0, a_1, \cdots, a_\nu)$ such that $h = H(\hat{a}_0, \cdots, \hat{a}_\nu)$. To prove that H is continuous, it suffices to prove that, if $m_i \in M, i = 1, 2, \cdots$ and $\hat{a}_k(m_i) \to \hat{a}_k(m^0)$, $k = 0, \cdots, \nu$, then $h(m_i) \to h(m^0)$. We may assume that $h(m_i)$ converges to a limit h'. But if m is a limit point of the sequence m_i, we have $\hat{a}_k(m) = \hat{a}_k(m^0)$, $k = 0, \cdots, \nu$, and $h' = h(m)$, so $h' = h(m^0)$.

The continuous function H on $\sigma(a_0, \cdots, a_\nu)$ satisfies the conditions

$$\sum_0^n z_k H^k = 0, \sum_0^n k z_k H^{k-1} \neq 0$$

in view of (3.2.3). From the implicit function theorem (Theorem 2.1.2) it follows that the first equation defines in some spherical neighborhood of any point in $\sigma(a_0, \cdots, a_\nu)$ an analytic function which coincides with the given function H in the intersection of $\sigma(a_0, \cdots, a_\nu)$ and the ball. Let 2δ be a positive lower bound for the radii of such balls; the existence is guaranteed by the compactness of $\sigma(a_0, \cdots, a_\nu)$. Then we obtain a uniquely defined analytic function H satisfying the equation $\Sigma_0^n z_k H^k = 0$ in the set of points at distance $< \delta$ from $\sigma(a_0, \cdots, a_\nu)$ and coinciding with the given function H on that set. In fact, if two balls with radius δ overlap, then the center of one is contained in the ball with radius 2δ and the same center as the other ball. This proves the lemma.

Proof of Theorem 3.2.5. Combination of Lemma 3.2.7 with Theorem 3.2.2 gives that there is an element $w_0 \in A$ such that $\hat{w}_0 = h$. This means that

$$\sum_0^n a_k {w_0}^k = b,$$

where $\hat{b} = \Sigma_0^n \hat{a}_k h^k = 0$, hence $\|b^j\|^{1/j} \to 0$ when $j \to \infty$ by Theorem 3.1.6. (We say that b belongs to the *radical* of B.) The other condition in (3.2.3) means that $\Sigma_1^n k\hat{a}_k {\hat{w}_0}^{k-1} \neq 0$ everywhere on M, that is, $\Sigma_1^n k a_k {w_0}^{k-1}$ is invertible. To solve equation (3.2.1) exactly, we now set $w = w_0 + u$. To guarantee that $\hat{w} = h = \hat{w}_0$, we must have u in the radical. The equation for u becomes

$$b + u \sum_1^n k a_k {w_0}^{k-1} + \cdots = 0$$

where dots indicate terms involving higher powers of u. The proof of Theorem 3.2.5 will therefore be completed by the following lemma.

Lemma 3.2.8. *In the equation*

$$\sum_0^n b_j u^j = 0$$

with coefficients in B, we assume that b_0 belongs to the radical and that b_1 is invertible. Then the equation has a solution u belonging to the radical.

Note that this is the special case of Theorem 3.2.5 with $h = 0$.

Proof. Since we can multiply by b_1^{-1}, it is no restriction to assume that $b_1 = e$. Now consider the equation

$$z_0 + w + z_2w^2 + \cdots + z_nw^n = 0,$$

where $z_0, z_2, \cdots, z_n$ are complex variables. By the implicit function theorem it has a unique analytic solution w in a neighborhood of 0 in $\mathbf{C}^n$, such that $w = 0$ when $z_0 = z_2 = \cdots = z_n = 0$, and we can write

$$w(z) = \sum_{\alpha \neq 0} c_\alpha z^\alpha.$$

Here $\alpha = (\alpha_0, \alpha_2, \cdots, \alpha_n)$ is a multiorder and

$$\sum |c_\alpha| r^{|\alpha|} < \infty$$

for some $r > 0$. We claim that $c_\alpha = 0$ except when $L(\alpha) = \alpha_0 - \alpha_2 - 2\alpha_3 - \cdots - (n-1)\alpha_n > 0$. Assume that this has already been proved for terms of degree $\leq k$ (it is obvious when $k = 0$ or 1, since $w = -z_0 +$ terms of higher order). Now if $\alpha^1, \cdots, \alpha^j$ are multiorders with $L(\alpha^i) > 0$, $i = 1, \cdots, j$, it follows that $L(\alpha^1 + \cdots + \alpha^j) \geq j$. Therefore terms of degree $\leq k + 1$ in $z_2w^2 + \cdots + z_nw^n$ are all of the form z^β with $L(\beta) \geq j - (j-1) = 1 > 0$, so that all terms of degree $k + 1$ in the expansion of w also satisfy this condition.

We shall now prove that the series $w(b)$ obtained by substituting b_j for z_j converges in the norm, which implies that $w(b)$ is in the radical since all terms in the series are radical elements. Choose $R > 1$ so that $\|b_j\| < R^{j-1}$, $2 \leq j \leq n$. Then we have

$$\|b^\alpha\| \leq \|b_0^{\alpha_0}\| \, \|b_2\|^{\alpha_2} \ldots \|b_n\|^{\alpha_n} \leq \|b_0^{\alpha_0}\| R^{\alpha_0} \quad \text{if } L(\alpha) > 0.$$

Since b_0 belongs to the radical, we have for sufficiently large α_0

$$\|b_0^{\alpha_0}\| R^{\alpha_0} \leq r^{|\alpha|}.$$

For we may assume that $r < 1$ and then we have $r^{|\alpha|} \geq r^{2\alpha_0}$, and $\|b_0^{\alpha_0}\|^{1/\alpha_0} \to 0$ when $\alpha_0 \to \infty$. Hence the series $w(b) = \Sigma\, c_\alpha b^\alpha$ is absolutely convergent. It is clear that $w(b)$ satisfies the desired equation, for $b_0 + w(b) + b_2 w(b)^2 + \cdots$ can be rearranged as a power series in b, all of whose coefficients are 0. The proof is complete.

Theorem 3.2.9. *Assume that M is not connected and write $M = M_0 \cup M_1$ where M_0 and M_1 are closed and disjoint. Then there exist two elements e_0 and $e_1 \in B$ such that $e_0 + e_1 = e$, $e_0 e_1 = 0$, $\hat{e}_0 = 1$ on M_0, and $\hat{e}_1 = 1$ on M_1.*

ITHACA COLLEGE LIBRARY

Proof. The function $h = 0$ on M_0, $h = 1$ on M_1, is continuous and satisfies the equation $h(1 - h) = 0$ which has only simple zeros. Hence we can find $e_1 \in B$ so that $e_1(e - e_1) = 0$ and $\hat{e}_1 = 0$ on M_0, $\hat{e}_1 = 1$ on M_1. If we take $e_0 = e - e_1$, the statements are valid.

The result means that e_0 and e_1 are idempotents with $e_0 e_1 = 0$ and $e_0 + e_1 = e$. Thus the ideals Be_0 and Be_1 are Banach algebras with units e_0 and e_1, respectively, and B is their direct sum. Conversely, it is clear that, if a Banach algebra can be decomposed into a direct sum, then the maximal ideal space is the union of the corresponding maximal ideal spaces, hence disconnected.

Theorem 3.2.10. *Let $a \in B$ be invertible and assume that there is a continuous function h on M with $h^n = \hat{a}$. Then there is an element $w \in A$ with $w^n = a$.*

Proof. That a is invertible means that $\hat{a} \neq 0$ everywhere on M, so h is a simple zero of $h^n - \hat{a}$.

Finally we shall give a theorem which shows that the Shilov boundary can be described by local conditions.

Theorem 3.2.11. *Let $m_0 \in M$ and assume that there is an open neighborhood V_0 of m_0 such that to every neighborhood V of m_0 there is an element $f \in B$ with*

$$\sup_{V_0 \setminus V} |\hat{f}| < \sup_{V} |\hat{f}|. \tag{3.2.4}$$

Then m_0 belongs to the Shilov boundary of M with respect to B.

In other words, m_0 is in the Shilov boundary if m_0 is locally in the Shilov boundary. The proof requires the following lemma.

Lemma 3.2.12. *Let $f_0 \in B$ and*

$$\{m; m \in M, \hat{f}_0(m) = 1\} = K' \cup K''$$

where K' and K'' are compact and disjoint. Then there are elements $f_1, \cdots, f_\nu$ in B and a function φ which is analytic in a neighborhood of $\sigma(f_0, \cdots, f_\nu)$ such that $\varphi(z)/(z_0 - 1)$ is analytic in a neighborhood of $\{(\hat{f}_0(m), \cdots, \hat{f}_\nu(m)); m \in K'\}$ but $\varphi(\hat{f}_0(m), \cdots, \hat{f}_\nu(m)) = 1$ when $m \in K''$.

Proof. We can choose $f_1, \cdots, f_n$ so that the map

$$M \ni m \to (\hat{f}_0(m), \cdots, \hat{f}_n(m)) \tag{3.2.5}$$

maps K' and K'' on disjoint (necessarily compact) sets K_f' and K_f''

Let $\chi \in C_0^\infty(\mathbf{C}^{n+1})$ be equal to 1 in a neighborhood of K_f'' and equal to 0 in a neighborhood of K_f'. We wish to take $\varphi(z) = \chi(z) + (z_0 - 1)v(z)$ where v is chosen so that φ is analytic, that is, $(1 - z_0)\bar{\partial}v = \bar{\partial}\chi$. In a neighborhood Ω of M_f (defined as the range of the mapping (3.2.5)), we have $(1 - z_0)^{-1}\bar{\partial}\chi \in C^\infty_{(0,1)}(\Omega)$ and this form is obviously $\bar{\partial}$ closed. By Lemma 3.2.3 we can choose $f_{n+1}, \cdots, f_\nu$ so that, if B' denotes the algebra generated by $f_0, \cdots, f_\nu$, the polynomially convex set $\sigma_{B'}(f_0, \cdots, f_\nu) \subset \mathbf{C}^{\nu+1}$ is mapped into Ω by the projection

$$\pi : (z_0, \cdots, z_\nu) \to (z_0, \cdots, z_n).$$

By Lemma 2.7.4 we can find a polynomial polyhedron contained in $\pi^{-1}\Omega$ and containing $\sigma_{B'}(f_0, \cdots, f_\nu)$, and if we apply Theorem 2.7.6 it now follows that there is a C^∞ function v in a neighborhood of $\sigma_{B'}(f_0, \cdots, f_\nu)$ such that $\varphi = \pi^*\chi + (z_0 - 1)v$ is analytic there. This proves the lemma.

Proof of Theorem 3.2.11. We can take the neighborhood V of m_0 open and relatively compact in V_0. According to the hypothesis we can choose $f_0 \in B$ so that

$$\sup_{V_0 \setminus V} |\hat{f}_0| < 1, \qquad \sup_V |\hat{f}_0| = 1. \tag{3.2.6}$$

We can normalize f_0 so that $\hat{f}_0(m) = 1$ for some $m \in V$. Now set

$$K' = \{m ; m \notin V_0, \hat{f}_0(m) = 1\},\ K'' = \{m ; m \in V_0, \hat{f}_0(m) = 1\} \subset\subset V,$$

and apply the lemma. For sufficiently small $\varepsilon > 0$, the function

$$\varphi_\varepsilon(z) = \varepsilon\varphi(z_0 - \varepsilon, z_1, \cdots, z_\nu)(z_0 - 1 - \varepsilon)^{-1}$$

is analytic in a neighborhood of $\sigma(f_0, \cdots, f_\nu)$ because $|\hat{f}_0(m)| \leq 1$ when $m \in V_0$, and $\varphi(z)/(z_0 - 1)$ is analytic in a neighborhood of the image of $M \setminus V_0$ under the map $m \to (\hat{f}_0(m), \cdots, \hat{f}_\nu(m))$. Hence it follows from Theorem 3.2.2 that there is an element $g_\varepsilon \in B$ such that $\hat{g}_\varepsilon = \varphi_\varepsilon(\hat{f}_0, \cdots, \hat{f}_\nu)$. In the complement of V we have $\hat{g}_\varepsilon = O(\varepsilon)$ when $\varepsilon \to 0$, but $\hat{g}_\varepsilon(m) = -1 + O(\varepsilon)$ when $\varepsilon \to 0$ if $m \in K''$. Hence $\complement V$ is not a boundary for any V, which proves that m_0 is in the Shilov boundary.

Notes. The origin of Theorem 3.2.2 is the theorem of Wiener and Lévy that, if f is a periodic function with absolutely convergent Fourier series and if F is analytic in the range of f, then $F(f)$ has an absolutely convergent Fourier series. In several variables such results were first given by Shilov [1] to whom many of the results of section 3.2 are due. With the exception of Theorem 3.2.11 which is due to Rossi [1], all the results of section 3.2 can be found in Arens and Calderón [1]. For many important algebras it is impossible to relax the condition in Theorem 3.2.2 that φ be analytic (see Helson, Kahane, Katznelson and Rudin [1]). For further applications of complex analysis in the theory of Banach algebras, we refer to a survey article by Royden [1].

Chapter IV

L^2 ESTIMATES AND EXISTENCE THEOREMS FOR THE $\bar{\partial}$ OPERATOR

Summary. In this chapter we abandon the classical methods for solving the Cousin problems (that is, solving the Cauchy–Riemann equations), of which an example was given in section 2.7. Instead we develop a technique for studying the Cauchy–Riemann equations where the main point is an L^2 estimate proved in sections 4.1 and 4.2. In sections 4.2 and 4.3 we deduce existence and approximation theorems for solutions of the Cauchy-Riemann equations in domains of holomorphy. At the same time a solution of the Levi problem is obtained.

One of the main advantages of the methods used here is that, without any additional effort, they give results on existence of solutions of the Cauchy–Riemann equations satisfying growth conditions. In section 4.4 we give some theorems of this kind which are important in the applications to differential equations in section 7.6. A special case is applied to prove a theorem on analytic functionals in section 4.5. This theorem plays the same role in a number of existence theorems for analytic functions as the Paley–Wiener theorem plays in existence theorems involving C^∞ functions. However, we do not develop the applications here, and section 4.5 is not needed for studying Chapters V–VII.

4.1. Preliminaries. Let Ω be an open set in $\mathbf{C}^n$. If φ is a continuous function in Ω, we denote by $L^2(\Omega,\varphi)$ the space of functions in Ω which are square integrable with respect to the measure $e^{-\varphi}\, d\lambda$, where $d\lambda$ is the Lebesgue measure. This is a subspace of the space $L^2(\Omega,\text{loc})$ of functions in Ω which are locally square integrable with respect to the Lebesgue measure, and it is clear that every function in $L^2(\Omega,\text{loc})$ belongs to $L^2(\Omega,\varphi)$ for some φ. By $L^2_{(p,q)}(\Omega,\varphi)$ we denote the space of forms of type (p,q) with coefficients in $L^2(\Omega,\varphi)$,

$$f = \sum_{|I|=p}{}' \sum_{|J|=q}{}' f_{I,J}\, dz^I \wedge d\bar{z}^J,$$

where Σ' means that the summation is performed only over strictly increasing multi-indices. We set

$$|f|^2 = \sum_{I,J}' |f_{I,J}|^2,$$

and

$$\|f\|_\varphi^2 = \int |f|^2 e^{-\varphi} \, d\lambda.$$

It is clear that $L^2(\Omega,\varphi)$ is a Hilbert space with this norm. Similarly we define $L^2_{(p,q)}(\Omega,\mathrm{loc})$ and $D_{(p,q)}(\Omega)$ where $D(\Omega)$ is a notation for $C_0^\infty(\Omega)$ which we sometimes use in order to avoid having too many subscripts and superscripts. The space $D_{(p,q)}(\Omega)$ is of course dense in $L^2_{(p,q)}(\Omega,\varphi)$ for every φ.

If φ_1 and φ_2 are two continuous functions in Ω, then the operator $\bar\partial$ defines a linear, closed, densely defined operator

$$T\colon L^2_{(p,q)}(\Omega,\varphi_1) \to L^2_{(p,q+1)}(\Omega,\varphi_2);$$

an element $u \in L^2_{(p,q)}(\Omega,\varphi_1)$ is in D_T if $\bar\partial u$, defined in the sense of distribution theory, belongs to $L^2_{(p,q+1)}(\Omega,\varphi_2)$, and then we set $Tu = \bar\partial u$. That T is closed follows from the fact that differentiation is a continuous operation in distribution theory, and the domain is dense since it contains $D_{(p,q)}(\Omega)$.

For suitable densities we want to prove that the range of T consists of all $f \in L^2_{(p,q+1)}(\Omega,\varphi_2)$ such that $\bar\partial f = 0$ (which is of course a necessary condition for f to be in the range of T). The following lemma reduces this question to the study of an estimate.

Lemma 4.1.1. *Let T be a linear, closed, densely defined operator from one Hilbert space H_1 to another H_2, and let F be a closed subspace of H_2 containing the range R_T of T. Then $F = R_T$ if and only if for some constant C*

$$\|f\|_{H_2} \le C\|T^*f\|_{H_1}, \qquad f \in F \cap D_{T^*}. \tag{4.1.1}$$

Proof. First assume that (4.1.1) is valid and let $g \in F$. Since $T^{**} = T$, the equation $Tu = g$ is equivalent to the identity

$$(u,T^*f)_{H_1} = (g,f)_{H_2}, \qquad f \in D_{T^*}.$$

If we prove that

$$|(g,f)_{H_2}| \le C\|g\|_{H_2}\|T^*f\|_{H_1}, \qquad f \in D_{T^*}, \tag{4.1.2}$$

an application of the Hahn–Banach theorem to the antilinear form $T^*f \to (g,f)_{H_2}$ will thus show that the equation $Tu = g$ has a solution u with

$$\|u\|_{H_1} \le C\|g\|_{H_2}. \tag{4.1.3}$$

(Since H_1 is a Hilbert space we could of course use a projection on R_{T^*} instead of the Hahn–Banach theorem.) To prove (4.1.2) we first note that, if f is orthogonal to F, we have $(g,f)_{H_2} = 0$, and $T^*f = 0$ since $R_T \subset F$. Hence it is sufficient to prove (4.1.2) when $f \in F \cap D_{T^*}$ and then it follows immediately from (4.1.1).

Conversely, assuming that $R_T = F$, we must prove that the set

$$B = \{f; f \in F \cap D_{T^*},\ \|T^*f\|_{H_1} \leq 1\}$$

is bounded. To do so it is sufficient to prove that B is weakly bounded in F, that is, that $|(f,g)_{H_2}|$ is bounded when $f \in B$ for every fixed $g \in F$. But by hypothesis we can choose $u \in D_T$ so that $Tu = g$, and this implies

$$|(f,g)_{H_2}| = |(T^*f,u)_{H_1}| \leq \|u\|_{H_1}, \qquad f \in B.$$

The lemma is proved.

In proving approximation theorems we shall need the information concerning the operator T^* which follows from (4.1.1).

Lemma 4.1.2. *Let T be a linear closed densely defined operator from one Hilbert space H_1 to another H_2, and let F be a closed subspace of H_2 containing the range R_T of T. Assume that* (4.1.1) *is valid. For every $v \in H_1$ which is orthogonal to the null space of T, one can then find $f \in D_{T^*}$ such that $T^*f = v$ and*

$$\|f\|_{H_2} \leq C\|v\|_{H_1}. \tag{4.1.4}$$

Proof. The null space of T is the space orthogonal to R_{T^*}, so the hypothesis means that v is in the closure of R_{T^*}. Now the orthogonal complement of F is in the null space of T^*, so the restriction of T^* to $F \cap D_{T^*}$ has the same range as T^*. But (4.1.1) shows that this restriction has a closed range. Hence we can find $f \in F \cap D_{T^*}$ so that $T^*f = v$, and (4.1.4) then follows from (4.1.1)

In our application of Lemma 4.1.1, the spaces H_1 and H_2 will be $L^2_{(p,q)}(\Omega,\varphi_1)$ and $L^2_{(p,q+1)}(\Omega,\varphi_2)$, respectively, T the operator between these spaces defined as explained above by the $\bar{\partial}$ operator, and F the set of all $f \in L^2_{(p,q+1)}(\Omega,\varphi_2)$ with $\bar{\partial}f = 0$ (in the sense of distribution theory). Let φ_3 be another continuous function and let S be the operator from $L^2_{(p,q+1)}(\Omega,\varphi_2)$ to $L^2_{(p,q+2)}(\Omega,\varphi_3)$ defined by $\bar{\partial}$. Then F is the null space of S, and to prove (4.1.1) it will be sufficient to show that

$$\|f\|^2_{\varphi_2} \leq C^2(\|T^*f\|^2_{\varphi_1} + \|Sf\|^2_{\varphi_3}), \qquad f \in D_{T^*} \cap D_S, \tag{4.1.5}$$

for the last term drops out when f is in the null space of S. If the densities are suitably chosen, it is enough to prove (4.1.5) when $f \in D_{(p,q+1)}(\Omega)$, for we have

Lemma 4.1.3. *Let η_ν, $\nu = 1, 2, \cdots$ be a sequence of functions in $C_0^\infty(\Omega)$ such that $0 \le \eta_\nu \le 1$ and $\eta_\nu = 1$ on any compact subset of Ω when ν is large. Suppose that $\varphi_2 \in C^1(\Omega)$ and that*

$$(4.1.6) \qquad e^{-\varphi_{j+1}} \sum_{k=1}^{n} |\partial \eta_\nu / \partial \bar{z}_k|^2 \le e^{-\varphi_j}; \qquad j = 1, 2; \ \nu = 1, 2, \cdots.$$

Then $D_{(p,q+1)}(\Omega)$ is dense in $D_{T^*} \cap D_S$ for the graph norm

$$f \to \|f\|_{\varphi_2} + \|T^*f\|_{\varphi_1} + \|Sf\|_{\varphi_3}.$$

Note that (4.1.6) means only a finite number of bounds for $\varphi_j - \varphi_{j+1}$ on each compact subset of Ω, so one can always find continuous functions $\varphi_1, \varphi_2, \varphi_3$ satisfying (4.1.6).

Proof of the lemma. Since

$$S(\eta_\nu f) - \eta_\nu Sf = \bar{\partial}\eta_\nu \wedge f, \qquad f \in D_S,$$

it follows from (4.1.6) that

$$|S(\eta_\nu f) - \eta_\nu Sf|^2 e^{-\varphi_3} \le |f|^2 e^{-\varphi_2}.$$

Hence the dominated convergence theorem gives

$$(4.1.7) \qquad \|S(\eta_\nu f) - \eta_\nu Sf\|_{\varphi_3} \to 0 \qquad \text{when } \nu \to \infty, \qquad f \in D_S.$$

If $f \in D_{T^*}$ and $\eta \in C_0^\infty(\Omega)$, it follows that $\eta f \in D_{T^*}$. In fact,

$$\begin{aligned}(\eta f, Tu)_{\varphi_2} &= (f, \bar{\eta} Tu)_{\varphi_2} = (f, T(\bar{\eta}u))_{\varphi_2} + (f, \bar{\eta}Tu - T(\bar{\eta}u))_{\varphi_2} \\ &= (\eta T^* f, u)_{\varphi_1} + (f, \bar{\eta}Tu - T(\bar{\eta}u))_{\varphi_2}, \qquad u \in D_T.\end{aligned}$$

Since no derivative of u occurs in the last term, it follows that $(\eta f, Tu)_{\varphi_2}$ is continuous for the norm $\|u\|_{\varphi_1}$, so there is an element $v \in L^2_{(p,q)}(\Omega, \varphi_1)$ with

$$(v, u)_{\varphi_1} = (\eta f, Tu)_{\varphi_2}, \qquad u \in D_T.$$

This means that $\eta f \in D_{T^*}$ and that $T^*(\eta f) = v$. When $\eta = \eta_\nu$, we obtain by estimating $\bar{\eta}_\nu Tu - T(\bar{\eta}_\nu u)$ as in the proof of (4.1.7) that

$$|(T^*(\eta_\nu f) - \eta_\nu T^* f, u)_{\varphi_1}| \le \int |f| e^{-\varphi_2/2} |u| e^{-\varphi_1/2} \, d\lambda,$$

which implies the bound

$$|T^*(\eta_\nu f) - \eta_\nu T^* f|^2 e^{-\varphi_1} \le |f|^2 e^{-\varphi_2}.$$

As above, we can therefore conclude by dominated convergence that

(4.1.8) $\|T^*(\eta_\nu f) - \eta_\nu T^* f\|_{\varphi_1} \to 0$ when $\nu \to \infty$ if $f \in D_{T^*}$.

Hence $\eta_\nu f \to f$ in the graph norm if $f \in D_{T^*} \cap D_S$.

To complete the proof we only have to approximate elements $f \in D_{T^*} \cap D_S$ with compact support in Ω by elements in $D_{(p,q+1)}(\Omega)$. This requires an elementary lemma:

Lemma 4.1.4. *Let χ be a function in $C_0^\infty(\mathbf{R}^N)$ with $\int \chi\, dx = 1$, and set $\chi_\varepsilon(x) = \varepsilon^{-N}\chi(x/\varepsilon)$, $x \in \mathbf{R}^N$. If $g \in L^2(\mathbf{R}^N)$ it follows that*

$$g * \chi_\varepsilon(x) = \int g(y)\chi_\varepsilon(x-y)\, dy = \int g(x - \varepsilon y)\chi(y)\, dy$$

is a C^∞ function such that $\|g * \chi_\varepsilon - g\|_{L_2} \to 0$ when $\varepsilon \to 0$. The support of $g * \chi_\varepsilon$ has no points at distance $> \varepsilon$ from the support of g if the support of χ lies in the unit ball.

Assuming the lemma for a moment we complete the proof of Lemma 4.1.3. If $f \in D_{T^*} \cap D_S$ has compact support, we define $f * \chi_\varepsilon$ by choosing χ as in Lemma 4.1.4 (with $N = 2n$) and letting the convolution act on each coefficient of f. The support of $f * \chi_\varepsilon$ is then contained in a fixed compact subset of Ω when $\varepsilon \to 0$, and the lemma gives that $\|f - f * \chi_\varepsilon\|_{\varphi_2} \to 0$. Since $S(f * \chi_\varepsilon) = (Sf) * \chi_\varepsilon$, we also obtain from the lemma that $\|Sf - S(f * \chi_\varepsilon)\|_{\varphi_3} \to 0$. The operator T^* does not have constant coefficients but we can write $e^{\varphi_2 - \varphi_1}T^* = \vartheta + a$ where ϑ is a constant coefficient differential operator and a is of degree 0. (Cf. formula (4.1.9) below.) Since

$$(\vartheta + a)(f * \chi_\varepsilon) = ((\vartheta + a)f) * \chi_\varepsilon + a(f * \chi_\varepsilon) - (af) * \chi_\varepsilon$$

and the right-hand side is L^2 convergent to the limit $(\vartheta + a)f + af - af$ according to Lemma 4.1.4, it follows that $\|T^*(f * \chi_\varepsilon) - T^* f\|_{\varphi_1} \to 0$, which completes the proof of Lemma 4.1.3.

Proof of Lemma 4.1.4. In the first integral defining $g * \chi_\varepsilon$ we can differentiate under the sign of integration any number of times to show that $g * \chi_\varepsilon \in C^\infty$. From the second expression for $g * \chi_\varepsilon$ and Minkowski's inequality, it follows that

$$\|g * \chi_\varepsilon\|_{L^2} \le C\|g\|_{L^2} \quad \text{where } C = \int |\chi|\, dx.$$

It is obvious that $g * \chi_\varepsilon - g \to 0$ uniformly if $g \in C_0^\infty(\mathbf{R}^N)$, which is a dense subset of L^2. Since the last statement in the lemma is also self-evident, we have $\|g * \chi_\varepsilon - g\|_{L^2} \to 0$ for all g in a dense set in L^2, and from the uniform bound just proved it follows that this is true for all $g \in L^2$. The proof is complete.

We conclude this section by actually computing T^*, which also gives another proof of (4.1.8). Thus choose

$$u = \sum_{|I|=p}' \sum_{|K|=q}' u_{I,K}\, dz^I \wedge d\bar{z}^K \in D_{(p,q)}(\Omega),$$

$$f = \sum_{|I|=p}' \sum_{|J|=q+1}' f_{I,J}\, dz^I \wedge d\bar{z}^J \in L^2_{(p,q+1)}(\Omega,\varphi_2).$$

Since $f_{I,J}$ is defined for all J as an antisymmetric function of the indices in J, and

$$\bar{\partial} u = \sum_{|I|=p}' \sum_{|K|=q}' \sum_{j=1}^{n} \partial u_{I,K}/\partial \bar{z}^j\, d\bar{z}^j \wedge dz^I \wedge d\bar{z}^K$$

we obtain, if $f \in D_{T^*}$,

$$\int \sum_{I,K}' (T^*f)_{I,K} \overline{u_{I,K}}\, e^{-\varphi_1}\, d\lambda = (T^*f,u)_{\varphi_1} = (f,Tu)_{\varphi_2}$$

$$= (-1)^p \int \sum_{I,K}' \sum_{j=1}^{n} f_{I,jK}\, \overline{\partial u_{I,K}/\partial \bar{z}_j}\, e^{-\varphi_2}\, d\lambda,$$

which means that

$$T^*f = (-1)^{p-1} \sum_{I,K}' \sum_{j=1}^{n} e^{\varphi_1}\, \partial(e^{-\varphi_2} f_{I,jK})/\partial z_j\, dz^I \wedge d\bar{z}^K. \tag{4.1.9}$$

4.2. Existence theorems in pseudoconvex domains. Choose a function $\psi \in C^\infty(\Omega)$ such that

$$\sum_{k=1}^{n} |\partial \eta_\nu/\partial \bar{z}_k|^2 \le e^\psi \text{ in } \Omega, \qquad \nu = 1, 2, \cdots. \tag{4.2.1}$$

If we set

$$\varphi_1 = \varphi - 2\psi,\ \varphi_2 = \varphi - \psi,\ \varphi_3 = \varphi, \tag{4.2.2}$$

the condition (4.1.6) is satisfied for any choice of φ. We shall now study $\|T^*f\|_{\varphi_1}$ and $\|Sf\|_{\varphi_3}$ when $f \in D_{(p,q+1)}(\Omega)$, keeping ψ fixed in all that follows but making all estimates uniform in φ so that we can make a suitable choice of φ at the end of the discussion. We assume that $\varphi \in C^2(\Omega)$.

First note that, since

$$\bar{\partial} f = \sum_{|I|=p}' \sum_{|J|=q+1}' \sum_{j=1}^{n} \partial f_{I,J}/\partial \bar{z}_j\, d\bar{z}_j \wedge dz^I \wedge d\bar{z}^J,$$

we obtain

$$|\bar{\partial} f|^2 = \sum_{I,J,L}' \sum_{j,l=1}^{n} \partial f_{I,J}/\partial \bar{z}_j \overline{\partial f_{I,L}/\partial \bar{z}_l} \varepsilon_{lL}^{jJ}$$

where $\varepsilon_{lL}^{jJ} = 0$ unless $j \notin J$, $l \notin L$, and $\{j\} \cup J = \{l\} \cup L$, in which case ε_{lL}^{jJ} is the sign of the permutation $\binom{jJ}{lL}$. We shall rearrange the terms in this sum. First consider the terms with $j = l$. Then we must have $J = L$ and $j \notin J$ if $\varepsilon_{lL}^{jJ} \neq 0$, so the sum of these terms is

$$\sum_{I,J}' \sum_{j \notin J} |\partial f_{I,J}/\partial \bar{z}_j|^2.$$

Next consider the terms with $j \neq l$. If $\varepsilon_{lL}^{jJ} \neq 0$, we must then have $l \in J$ and $j \in L$, and deletion of l from J or j from L gives the same multi-index K. Since

$$\varepsilon_{lL}^{jJ} = \varepsilon_{jlK}^{jJ} \varepsilon_{ljK}^{jlK} \varepsilon_{lL}^{ljK} = -\varepsilon_{lK}^{J} \varepsilon_{L}^{jK},$$

the sum of the terms in question is

$$-\sum_{I,K}' \sum_{j \neq l} \partial f_{I,lK}/\partial \bar{z}_j \overline{\partial f_{I,jK}/\partial \bar{z}_l}.$$

Hence we obtain

(4.2.3) $$|\bar{\partial} f|^2 = \sum_{I,J}' \sum_{j} |\partial f_{I,J}/\partial \bar{z}_j|^2 - \sum_{I,K}' \sum_{j,k} \partial f_{I,jK}/\partial \bar{z}_k \overline{\partial f_{I,kK}/\partial \bar{z}_j}.$$

(When $p = 0$, $q = 1$, this follows from the fact that

$$|\bar{\partial} f|^2 = \sum |\partial f_j/\partial \bar{z}_k - \partial f_k/\partial \bar{z}_j|^2/2.)$$

Next we consider T^*f. With the notation

(4.2.4) $$\delta_j w = e^{\varphi}\, \partial(w e^{-\varphi})/\partial z_j = \partial w/\partial z_j - w \partial \varphi/\partial z_j,$$

we obtain from (4.1.9)

$$e^{\psi}\, T^*f = (-1)^{p-1} \sum_{I,K}' \sum_{j=1}^{n} \delta_j f_{I,jK}\, dz^I \wedge d\bar{z}^K$$

$$+ (-1)^{p-1} \sum_{I,K}' \sum_{j=1}^{n} f_{I,jK}\, \partial \psi/\partial z_j\, dz^I \wedge d\bar{z}^K.$$

Hence

$$\int \sum_{I,K}' \sum_{j,k=1}^{n} \delta_j f_{I,jK} \overline{\delta_k f_{I,kK}}\, e^{-\varphi}\, d\lambda \leq 2\|T^*f\|_{\varphi_1}^2 + 2\int |f|^2 |\partial \psi|^2 e^{-\varphi}\, d\lambda.$$

Combining this estimate with (4.2.3), we obtain

$$\int \sum_{I,K}' \sum_{j,k=1}^{n} (\delta_j f_{I,jK} \overline{\delta_k f_{I,kK}} - \partial f_{I,jK}/\partial \bar{z}_k \overline{\partial f_{I,kK}/\partial \bar{z}_j})\, e^{-\varphi}\, d\lambda \tag{4.2.5}$$

$$+ \int \sum_{I,J}' \sum_{j=1}^{n} |\partial f_{I,J}/\partial \bar{z}_j|^2 e^{-\varphi}\, d\lambda \le 2\|T^*f\|_{\varphi_1}^2 + \|Sf\|_{\varphi_3}^2 + 2\int |f|^2 |\partial\psi|^2 e^{-\varphi}\, d\lambda.$$

Now the operators $\partial/\partial \bar{z}_k$ and $-\delta_k$ are adjoint in the sense that

$$\int w_1 \overline{\partial w_2/\partial \bar{z}_k}\, e^{-\varphi}\, d\lambda = -\int \delta_k w_1 \overline{w_2}\, e^{-\varphi}\, d\lambda, \qquad w_1, w_2 \in C_0^\infty(\Omega);$$

and we have the commutation relations

$$\delta_j \partial/\partial \bar{z}_k - \partial/\partial \bar{z}_k \delta_j = \partial^2\varphi/\partial \bar{z}_k \partial z_j. \tag{4.2.6}$$

Shifting the differentiations to the left in the first sum in (4.2.5) therefore gives

$$\sum_{I,K}' \int \sum_{j,k=1}^{n} f_{I,jK} \overline{f_{I,kK}}\, \partial^2\varphi/\partial z_j \partial \bar{z}_k\, e^{-\varphi}\, d\lambda + \sum_{I,J}' \sum_{j=1}^{n} \int |\partial f_{I,J}/\partial \bar{z}_j|^2 e^{-\varphi}\, d\lambda \tag{4.2.7}$$

$$\le 2\|T^*f\|_{\varphi_1}^2 + \|Sf\|_{\varphi_3}^2 + 2\int |f|^2 |\partial\psi|^2 e^{-\varphi}\, d\lambda, \qquad f \in D_{(p,q+1)}(\Omega).$$

Now assume that the function φ is strictly plurisubharmonic,

$$\sum_{j,k=1}^{n} \partial^2\varphi/\partial z_j \partial \bar{z}_k\, w_j \bar{w}_k \ge c \sum_{1}^{n} |w_j|^2, \qquad w \in \mathbf{C}^n, \tag{4.2.8}$$

where c is a positive continuous function in Ω. Then it follows from (4.2.7) that

$$\int (c - 2|\partial\psi|^2)|f|^2 e^{-\varphi}\, d\lambda \le 2\|T^*f\|_{\varphi_1}^2 + \|Sf\|_{\varphi_3}^2, \qquad f \in D_{(p,q+1)}(\Omega). \tag{4.2.9}$$

Recalling Lemma 4.1.3, we have now proved

Lemma 4.2.1. *With* $\varphi_1, \varphi_2, \varphi_3$ *defined by* (4.2.2), *where* $\varphi, \psi \in C^2(\Omega)$, *we have*

$$\|f\|_{\varphi_2}^2 \le \|T^*f\|_{\varphi_1}^2 + \|Sf\|_{\varphi_3}^2, \qquad f \in D_{T^*} \cap D_S, \tag{4.2.10}$$

provided that

$$\sum_{j,k=1}^{n} \partial^2\varphi/\partial z_j \partial \bar{z}_k\, w_j \bar{w}_k \ge 2(|\bar\partial\psi|^2 + e^{\psi}) \sum_{1}^{n} |w_j|^2, \qquad w \in \mathbf{C}^n. \tag{4.2.11}$$

We can now easily prove an existence theorem.

Theorem 4.2.2. *Let* Ω *be a pseudoconvex open set in* $\mathbf{C}^n$. *Then the equation* $\bar\partial u = f$ *has (in the sense of distribution theory) a solution* $u \in L^2_{(p,q)}(\Omega, \mathrm{loc})$ *for every* $f \in L^2_{(p,q+1)}(\Omega, \mathrm{loc})$ *such that* $\bar\partial f = 0$.

Proof. In view of Theorem 2.6.11, we can choose a strictly plurisubharmonic function $p \in \mathbf{C}^\infty(\Omega)$ such that

$$K_c = \{z\,; z \in \Omega, p(z) \leq c\} \subset \subset \Omega, \quad \text{for every } c \in \mathbf{R}.$$

Let

$$\sum_{j,k=1}^{n} \partial^2 p/\partial z_j \partial \bar{z}_k \, w_j \bar{w}_k \geq m \sum_{1}^{n} |w_j|^2,$$

where $0 < m \in C^0(\Omega)$. If χ is a C^∞ *convex* increasing function and $\varphi = \chi(p)$, we obtain

$$\sum_{j,k=1}^{n} \partial^2 \varphi/\partial z_j \partial \bar{z}_k \, w_j \bar{w}_k \geq \chi'(p) m \sum_{1}^{n} |w_j|^2.$$

Hence φ satisfies (4.2.11) if

$$\chi'(p) m \geq 2(|\bar{\partial}\psi|^2 + e^\psi),$$

that is, if

$$\chi'(t) \geq \sup_{K_t} 2(|\bar{\partial}\psi|^2 + e^\psi)/m. \tag{4.2.12}$$

The right-hand side of (4.2.12) is a finite increasing function of t, which is defined when $t \geq \min p$. Hence there exist increasing C^∞ functions χ' satisfying (4.2.12). It is clear that we can choose χ so that in addition any given $f \in L^2_{(p,q+1)}(\Omega,\text{loc})$ belongs to $L^2_{(p,q+1)}(\Omega,\varphi - \psi)$. But then it follows from Lemma 4.1.1 that the equation $\bar{\partial}u = f$ has a solution $u \in L^2_{(p,q)}(\Omega,\varphi - 2\psi)$. This proves the theorem.

We shall now examine the regularity properties of the solution u of the equation $\bar{\partial}u = f$ which we have obtained. In doing so it is important to note that the solution of the equation $Tu = f$ given by Lemma 4.1.1 can be chosen orthogonal to the null space of T, that is, in (the closure of) the range of T^*. This will yield an additional differential equation for u which is essential in proving the smoothness of u.

Let W^s, where s is a non-negative integer, denote the space of functions in $\mathbf{C}^n$ whose derivatives of order $\leq s$ are in L^2. By $W^s(\Omega,\text{loc})$ we denote the set of functions in Ω satisfying the same condition on compact subsets of Ω. The space of forms of type (p,q) with coefficients in this space is accordingly denoted by $W^s_{(p,q)}(\Omega,\text{loc})$. If f is of type $(p,q+1)$ $(p,q \geq 0)$, we set

$$\vartheta f = \sum_{I,K}{}' \sum_{j=1}^{n} \partial f_{I,jK}/\partial z_j \, dz^I \wedge d\bar{z}^K.$$

This is essentially the principal part of the differential operator in (4.1.9).

Lemma 4.2.3. *If $f \in L^2_{(p,q+1)}(\mathbf{C}^n)$ has compact support, if $\bar\partial f \in L^2_{(p,q+2)}(\mathbf{C}^n)$ and $\vartheta u \in L^2_{(p,q)}(\mathbf{C}^n)$, then $u \in W^1_{(p,q+1)}(\mathbf{C}^n)$.*

Proof. First note that if $f \in D_{(p,q+1)}$, then (4.2.7) with $\varphi = \psi = 0$ gives

$$\sum_{I,J}{}' \sum_{j=1}^{n} \int |\partial f_{I,J}/\partial \bar z_j|^2 \, d\lambda \le 2\|\vartheta f\|_0{}^2 + \|\bar\partial f\|_0{}^2. \tag{4.2.13}$$

If f only satisfies the hypotheses in the lemma, we can form a regularization $f * \chi_\varepsilon$ of f as defined in Lemma 4.1.4. If we apply (4.2.13) to $f * \chi_\varepsilon - f * \chi_\delta$, noting that $\vartheta(f * \chi_\varepsilon) = (\vartheta f) * \chi_\varepsilon \to \vartheta f$ in $L^2_{(p,q)}(\mathbf{C}^n)$ and the corresponding fact for $\bar\partial(f * \chi_\varepsilon)$, it follows that $\chi_\varepsilon * \partial f_{I,J}/\partial \bar z_j$ converges in L^2 for all I, J, j when $\varepsilon \to 0$. Hence $\partial f_{I,J}/\partial \bar z_j \in L^2$. The proof is therefore reduced to the following lemma.

Lemma 4.2.4. *If $w \in L^2(\mathbf{C}^n)$ has compact support and $\partial w/\partial \bar z_j \in L^2$ for $j = 1, \cdots, n$, then $w \in W^1$.*

Proof. We only have to prove that $\partial w/\partial z_j \in L^2$. If $w \in C_0{}^\infty$, two integrations by parts give

$$\int |\partial w/\partial z_j|^2 \, d\lambda = \int \partial w/\partial z_j \, \overline{\partial w/\partial z_j} \, d\lambda = -\int \partial^2 w/\partial \bar z_j \partial z_j \, \bar w \, d\lambda = \int |\partial w/\partial \bar z_j|^2 \, d\lambda.$$

Using this result in the same way as we used (4.2.13) in the proof of the previous lemma, we obtain that $\partial w/\partial z_j \in L^2$.

We can now give an improvement of Theorem 4.2.2.

Theorem 4.2.5. *Let Ω be a pseudoconvex open set in $\mathbf{C}^n$, and let $0 \le s \le \infty$. Then the equation $\bar\partial u = f$ has a solution $u \in W^{s+1}_{(p,q)}(\Omega,\text{loc})$ for every $f \in W^s_{(p,q+1)}(\Omega,\text{loc})$ such that $\bar\partial f = 0$. Every solution of the equation $\bar\partial u = f$ has this property when $q = 0$.*

Proof. (a) First assume that $q = 0$. We know from Theorem 4.2.2 that the equation $\bar\partial u = f$ has a solution $u = \Sigma' u_I \, dz^I \in L^2_{(p,0)}(\Omega,\text{loc})$. The equation $\bar\partial u = f$ means that

$$\partial u_I/\partial \bar z_j = f_{I,j} \in W^s(\Omega,\text{loc})$$

for all I and j. Suppose that $u \in W^\sigma(\Omega,\text{loc})$ for a certain finite σ with $0 \le \sigma \le s$; we know that this is true if $\sigma = 0$. If $\chi \in C_0{}^\infty(\Omega)$ we then obtain

$$\partial(\chi u_I)/\partial \bar z_j = \chi f_{I,j} + \partial\chi/\partial \bar z_j \, u_I \in W^\sigma.$$

If v is a derivative of order σ of χu_I, it follows that $\partial v/\partial \bar z_j \in L^2$ for every j. Hence $v \in W^1$ by Lemma 4.2.4, that is, all derivatives of χu_I of order $\sigma + 1$ are in L^2. This means that $u_I \in W^{\sigma+1}(\Omega,\text{loc})$. Repeating the argument, wc conclude that $u_I \in W^{s+1}(\Omega,\text{loc})$.

(b) Next assume that $q > 0$. As pointed out after the proof of Theorem 4.2.2, the solution of the equation $\bar{\partial}u = f$ given in that theorem can be chosen in (the closure of) the range of T^*. In view of (4.1.9) and the fact that $\vartheta^2 = 0$, we have

$$\vartheta\,(e^{-\varphi_1}u) = 0, \qquad \bar{\partial}u = f.$$

This can also be written

$$\bar{\partial}u = f, \qquad \vartheta u = au,$$

where a is a differential operator of order 0 with C^∞ coefficients acting on u. Assume that we have already proved that $u \in W^{\sigma}_{(p,q)}(\Omega,\text{loc})$ for a certain finite σ with $0 \leq \sigma \leq s$. If $\chi \in C_0{}^\infty(\Omega)$, we obtain

$$\bar{\partial}(\chi u) \in W^{\sigma}_{(p,q+1)}, \qquad \vartheta(\chi u) \in W^{\sigma}_{(p,q-1)}.$$

If D is a differentiation of order σ, the form $D\,(\chi u)$ satisfies the hypotheses of Lemma 4.2.3, which proves that $D\,(\chi u) \in W^1$. Hence $\chi u \in W^{\sigma+1}_{(p,q)}$, that is, $u \in W^{\sigma+1}_{(p,q)}(\Omega,\text{loc})$. This completes the proof.

Corollary 4.2.6. *If Ω is pseudoconvex, the equation $\bar{\partial}u = f$ has a solution $u \in C^\infty_{(p,q)}(\Omega)$ for every $f \in C^\infty_{(p,q+1)}(\Omega)$ such that $\bar{\partial}f = 0$.*

Proof. By the well-known Sobolev lemma, we have

$$W^{s+2n}_{(p,q)}(\Omega,\text{loc}) \subset C^s_{(p,q)}(\Omega) \tag{4.2.14}$$

so the corollary follows from Theorem 4.2.5. (An elementary proof of (4.2.14) is obtained from the trivial fact that

$$\sup |w| \leq \int |\partial^{2n}w/\partial x_1 \cdots \partial x_{2n}|\, d\lambda, \qquad w \in C_0{}^\infty(\mathbf{C}^n),$$

for as in the proof of Lemma 4.2.3 one obtains from this inequality that every $w \in W^{2n}$ with compact support is a continuous function after a change of its definition on a null set. The details are left to the reader.)

Combining Corollary 4.2.6 with Theorem 2.7.10, we obtain

Theorem 4.2.7. *If Ω is a pseudoconvex open set in $\mathbf{C}^n$, then $H^r(\Omega,\mathbf{C}) = 0$ when $r > n$.*

Remark. One can prove more by Morse theory, namely that $H^r(\Omega,\mathbf{Z}) = 0$ for $r > n$ (see Milnor [1]). On the other hand, $H^r(\Omega,\mathbf{C})$ need not vanish when $r \leq n$. For let

$$\Omega = \{z\,;\, z \in \mathbf{C}^n,\ \tfrac{1}{2} < |z_j| < 2, j = 1, \cdots, n\}.$$

This is a domain of holomorphy since it is the product of open sets in **C** (which are thus domains of holomorphy). Hence Ω is pseudoconvex. The form $f = dz_1 \wedge \cdots \wedge dz_r/z_1 \cdots z_r$, where $1 \le r \le n$, is closed, but if γ is the closed cycle $z_1 = e^{i\theta_1}, \cdots, z_r = e^{i\theta_r}$, $(\theta_j \in \mathbf{R})$, $z_{r+1} = \cdots = z_n = 1$, then the integral of f over γ is equal to $(2\pi i)^r \neq 0$. Hence f is not the differential of any form of degree $r - 1$. (See also an example in section 2.7.)

We can now prove a converse of Theorem 2.6.5.

Theorem 4.2.8. *An open set in* $\mathbf{C}^n$ *is a domain of holomorphy if (and by Theorem 2.6.5 only if) it is pseudoconvex.*

This is a consequence of Corollary 4.2.6 and the following

Theorem 4.2.9. *Let* Ω *be an open set in* $\mathbf{C}^n$ *such that the equation* $\bar\partial u = f$ *has a solution* $u \in C^\infty_{(0,q)}(\Omega)$ *for every* $f \in C^\infty_{(0,q+1)}(\Omega)$ *such that* $\bar\partial f = 0$ $(q = 0, \cdots, n-2)$. *Then* Ω *is a domain of holomorphy.*

Proof. The theorem is true when $n = 1$, since every open set in **C** is a domain of holomorphy. We shall prove it by induction over the dimension n, so we assume that it has already been proved for $n - 1$ dimensions.

It is enough to prove that for every open convex set $D \subset \Omega$, such that some boundary point z^0 of D is on the boundary $\partial\Omega$ of Ω, there is an analytic function in Ω which cannot be continued analytically to a neighborhood of z^0. We may assume that the coordinates are chosen so that $z^0 = 0$ and the plane $z_n = 0$ has a non-empty intersection D_0 with D. Then the convexity of D shows that 0 is on the boundary of D_0 and therefore on the boundary of

$$\omega = \{z\,;\, z \in \Omega, z_n = 0\}.$$

(We can regard ω as an open set in $\mathbf{C}^{n-1}$.) Let j be the natural injection of ω into Ω, and let π be the projection of $\mathbf{C}^n$ on $\mathbf{C}^{n-1}$, obtained by dropping the last coordinate. The main part of the proof is now to show that for every $f \in C^\infty_{(0,q)}(\omega) (q \ge 0)$ with $\bar\partial f = 0$, one can find $F \in C^\infty_{(0,q)}(\Omega)$ such that $\bar\partial F = 0$ and $f = j^*F$. (Cf. Lemma 2.7.5. The notation j^* was introduced in section 2.1.) To construct F we note that the sets ω and $M = \{z\,;\, z \in \Omega, \pi z \notin \omega\}$ are disjoint and relatively closed in Ω, so one can find $\varphi \in C^\infty(\Omega)$ such that $\varphi = 1$ in a neighborhood of ω and $\varphi = 0$ in a neighborhood of M. The form $\varphi\pi^*f$, defined as 0 where $\varphi = 0$, is then in $C^\infty_{(0,q)}(\Omega)$, and $j^*\varphi\pi^*f = f$. We now set

$$F = \varphi\pi^*f - z_n v$$

with $v \in C^\infty_{(0,q)}(\Omega)$ to be chosen so that $\bar\partial F = 0$. This means that

$$\bar\partial v = z_n^{-1} \bar\partial\varphi \wedge \pi^* f$$

and, since the right-hand side is in $C^\infty_{(0,q+1)}(\Omega)$ and is $\bar\partial$ closed, the existence of v follows from the assumption in the theorem. We have $j^*F = j^*\pi^*f = f$, so F has the required properties.

We can now prove that the hypotheses in the theorem are fulfilled if Ω is replaced by ω. In fact, for given $f \in C^\infty_{(0,q+1)}(\omega)$ with $\bar\partial f = 0$, we have proved that there is a form $F \in C^\infty_{(0,q+1)}(\Omega)$ with $\bar\partial F = 0$ and $j^*F = f$. By hypothesis the equation $\bar\partial U = F$ has a solution $U \in C^\infty_{(0,q)}(\Omega)$. Setting $u = j^*U$, we obtain

$$\bar\partial u = j^*\bar\partial U = j^*F = f.$$

By the inductive hypothesis it follows that ω is a domain of holomorphy, so there is a function f which is analytic in ω but cannot be continued analytically to a neighborhood of $\bar D_0$. If we choose F analytic in Ω so that $j^*F = f$, that is, $F = f$ in ω, it follows that F cannot be continued to a neighborhood of $\bar D$. Hence Ω is a domain of holomorphy, which completes the proof.

4.3. Approximation theorems. We shall now show that the L^2 estimates proved in section 4.2 also give approximation theorems. The essential step is the following

Lemma 4.3.1. *Let p be a strictly plurisubharmonic C^∞ function in Ω such that*

$$K_c = \{z\,;\, z \in \Omega,\ p(z) \le c\} \subset\subset \Omega \quad \text{for every } c \in \mathbf{R}.$$

Every function which is analytic in a neighborhood of K_0 can then be approximated in L^2 norm over K_0 by functions in $A(\Omega)$.

Proof. By the Hahn–Banach theorem it suffices to prove that, if $v \in L^2(K_0)$ and if

$$\int_{K_0} u\,\bar v\, d\lambda = 0 \tag{4.3.1}$$

for every $u \in A(\Omega)$, then (4.3.1) is true for every u which is analytic only in a neighborhood of K_0. Extend the definition of v by setting $v = 0$ outside K_0. Then (4.3.1) implies that $v\,e^{\varphi_1}$ is orthogonal to the null space N_T of the operator T discussed in the previous section (with $p = q = 0$), for N_T is a subspace of $A(\Omega)$. (All elements in the null space are C^∞ functions by Theorem 4.2.5.) When the estimate (4.2.10) is valid, it follows from Lemma 4.1.2 that one can find $f = \Sigma_1^n f_j d\bar{z}_j \in D_{T^*}$ so that

$$\|f\|_{\varphi_2} \leq \|e^{\varphi_1} v\|_{\varphi_1}$$

and $T^*f = v\, e^{\varphi_1}$. By (4.1.9) this equation implies that

$$v\, e^{\varphi_1} = -e^{\varphi_1} \sum_{j=1}^{n} \partial(e^{-\varphi_2} f_j)/\partial z_j$$

in the sense of distribution theory. Writing $g = f\, e^{-\varphi_2}$ we therefore have $v = -\Sigma_1^n\, \partial g_j/\partial z_j$ and

$$\int_\Omega \sum_1^n |g_j|^2\, e^{\varphi_2}\, d\lambda \leq \int_\Omega |v|^2\, e^{\varphi_1}\, d\lambda, \tag{4.3.2}$$

still under the assumption that (4.2.10) is valid.

In the proof of Theorem 4.2.2 we found that (4.2.10) is valid if $\varphi_1, \varphi_2, \varphi_3$ are defined by (4.2.2) with $\varphi = \chi(p)$, where χ is convex and satisfies (4.2.12). Choose a sequence χ_ν of convex functions satisfying (4.2.12), so that $\chi_\nu(t)$ is independent of ν if $t \leq 0$ but $\chi_\nu \nearrow \infty$ when $\nu \to \infty$ if $t > 0$. We can then for every ν choose g_j^ν, $j = 1, \cdots, n$, so that

$$v = -\sum_1^n \partial g_j^\nu/\partial z_j, \tag{4.3.3}$$

and

$$\int \sum_1^n |g_j^\nu|^2 \exp(\chi_\nu(p) - \psi)\, d\lambda \leq C \tag{4.3.4}$$

for a certain constant C which does not depend on ν, for the right-hand side of (4.3.2) only involves integration over K_0 and $\chi_\nu(p)$ is independent of ν there. Now $\chi_1 \leq \chi_\nu$, so we can choose a subsequence of the sequence g^ν which converges weakly in the Hilbert space $L^2_{(0,1)}(\Omega, \psi - \chi_1(p))$ to a limit g. From (4.3.4) we obtain that $g = 0$ where $p > 0$, and (4.3.3) gives that $v = -\Sigma_1^n\, \partial g_j/\partial z_j$ in the distribution sense, that is,

$$\int u\, \bar{v}\, d\lambda = \int \sum_1^n \partial u/\partial \bar{z}_j\, \bar{g}_j\, d\lambda \tag{4.3.5}$$

for all $u \in C_0^\infty(\Omega)$. In fact, this follows from (4.3.3) with g replaced by g^ν. Since v and g vanish outside K_0, the identity (4.3.5) is valid for every u which is in C^∞ only in a neighborhood of K_0, and if u is analytic in a neighborhood of K_0 it follows that (4.3.1) holds. The proof is complete.

The following reformulation of Lemma 4.3.1 is more useful.

Theorem 4.3.2. *Let Ω be a pseudoconvex open set in $\mathbf{C}^n$ and K a compact subset of Ω such that $\hat{K}_\Omega{}^P = K$ (see Definition 2.6.6). Every function which is analytic in a neighborhood of K can then be approximated uniformly on K by functions in $A(\Omega)$.*

Proof. Let u be analytic in a neighborhood ω of K. By Theorem 2.6.11 we can choose a strictly plurisubharmonic function $p \in C^\infty(\Omega)$ such that p satisfies the hypotheses in Lemma 4.3.1 and K is in the interior of K_0, which is itself in the interior of ω. According to Lemma 4.3.1 there is a sequence $u_j \in A(\Omega)$ such that $u_j - u \to 0$ in $L^2(K_0)$. By Theorem 2.2.3 this implies that $u_j - u \to 0$ uniformly on K. The proof is complete.

Note that since $\hat{K}_\Omega{}^P \subset \hat{K}_\Omega$ (the $A(\Omega)$-hull of K, see Definition 2.5.2), Theorem 4.3.2 is also valid if we replace $\hat{K}_\Omega{}^P$ by $\hat{K}_\Omega$. We can therefore give a result parallel to Theorem 1.3.4 which extends Theorem 2.7.3.

Theorem 4.3.3. *Let $\Omega_1 \subset \Omega_2$ be domains of holomorphy. Then the following conditions are equivalent:*

(i) *Every function in $A(\Omega_1)$ can be approximated by functions in $A(\Omega_2)$ uniformly on every compact subset of Ω_1. (Ω_1 is then called a Runge domain relative to Ω_2.)*
(ii) *For every compact set $K \subset \Omega_1$ we have $\hat{K}_{\Omega_2} = \hat{K}_{\Omega_1}$.*
(iii) *For every compact set $K \subset \Omega_1$ we have $\hat{K}_{\Omega_2} \cap \Omega_1 = \hat{K}_{\Omega_1}$.*
(iv) *For every compact set $K \subset \Omega_1$ we have $\hat{K}_{\Omega_2} \cap \Omega_1 \subset\subset \Omega_1$.*

Proof. It is obvious that (i) $\Rightarrow$ (iii) $\Rightarrow$ (iv) (for Ω_1 is assumed to be a domain of holomorphy). If we set $K' = \hat{K}_{\Omega_2} \cap \Omega_1$ and $K'' = \hat{K}_{\Omega_2} \cap \complement\Omega_1$, then (iv) implies that the disjoint sets K' and K'' are compact, for $\hat{K}_{\Omega_2}$ is compact since Ω_2 is a domain of holomorphy. For every $f \in A(\Omega_1)$ it follows from Theorem 4.3.2 that the function which is equal to f on K' and equal to 1 on K'' can be approximated uniformly on $\hat{K}_{\Omega_2} = K' \cup K''$ by functions in $A(\Omega_2)$. This proves (i), and choosing $f = 0$ we obtain $K'' = \emptyset$. Hence (iv) (and the equivalent condition (iii)) implies (ii) and since the opposite implication is obvious, the theorem is proved.

We shall finally give a strengthened form of Theorem 4.2.8.

Theorem 4.3.4. *Let K be a compact subset of a pseudoconvex open set $\Omega \subset \mathbf{C}^n$. Then $\hat{K}_\Omega$ (Definition 2.5.2) is equal to $\hat{K}_\Omega{}^P$ (Definition 2.6.6).*

This of course contains Theorem 4.2.8 and we shall prove Theorem 4.3.4 using that result. For a direct argument, see also the proof of Theorem 5.2.10.

Proof of Theorem 4.3.4. It is trivial that $\hat{K}_\Omega{}^P \subset \hat{K}_\Omega$. To prove the opposite inclusion, let ω be any open neighborhood of $\hat{K}_\Omega{}^P$. Using Theorem 2.6.11 we can choose a continuous plurisubharmonic function p in Ω such that $p < 0$ in K, but

$$\Omega_0 = \{z;\, z \in \Omega,\, p(z) < 0\}$$

is contained in ω. Since Ω_0 is obviously pseudoconvex, it follows from Theorem 4.2.8 that Ω and Ω_0 are domains of holomorphy. Moreover, Ω_0 is a Runge domain relative to Ω, for every function in $A(\Omega_0)$ can be uniformly approximated by functions in $A(\Omega)$ on the compact sets $\{z;\, z \in \Omega,\, p(z) \leq c\}$, $c < 0$, in view of Theorem 4.3.2. Hence condition (ii) in Theorem 4.3.3 shows that $\hat{K}_\Omega \subset \Omega_0 \subset \omega$. Thus $\hat{K}_\Omega \subset \hat{K}_\Omega{}^P$ which completes the proof.

4.4. Existence theorems in L^2 spaces. In sections 4.2 and 4.3 we have used the L^2 estimates to study the $\bar\partial$ operator in C^∞. However, more precise global information is obtained if one keeps the L^2 norms as we shall do in this section.

Lemma 4.4.1. *Let Ω be a pseudoconvex open set in $\mathbf{C}^n$ and let φ be a real valued function in $C^2(\Omega)$ such that*

$$c\sum_1^n |w_j|^2 \leqq \sum_{j,k=1}^n \partial^2\varphi/\partial z_j \partial \bar{z}_k w_j \bar{w}_k; \qquad z \in \Omega,\ w \in \mathbf{C}^n, \tag{4.4.1}$$

where c is a positive continuous function in Ω. If $g \in L^2_{(p,q+1)}(\Omega, \varphi)$ and $\bar\partial g = 0$, it follows that one can find $u \in L^2_{(p,q)}(\Omega, \varphi)$ with $\bar\partial u = g$ and

$$\int |u|^2 e^{-\varphi} d\lambda \leqq 2\int |g|^2 e^{-\varphi}/c\, d\lambda, \tag{4.4.2}$$

provided that the right hand side is finite.

Proof. Using Theorem 2.6.11 we choose a C^∞ strictly plurisubharmonic function s in Ω such that for every real a

$$\Omega_a = \{z \in \Omega;\, s(z) < a\} \Subset \Omega.$$

For any a the cutoff functions η_ν used in section 4.2 can be chosen equal to 1 in Ω_{a+1}, and we can then find $\psi \geqq 0$ satisfying (4.2.1) so that $\psi = 0$ in Ω_{a+1}. We replace φ by $\varphi' = \varphi + \chi(s)$ where $\chi \geqq 0$ is equal to 0 in $(-\infty, a)$ and χ is convex and so rapidly increasing that

$$\varphi'-2\psi = \varphi+\chi(s)-2\psi \geqq \varphi,\ \chi'(s)\sum_{j,k=1}^{n} \partial^2 s/\partial z_j\partial\bar{z}_k w_j\bar{w}_k \geqq 2|\partial\psi|^2 \sum_1^n |w_j|^2.$$

If we apply (4.2.9) with φ replaced by φ' and $\varphi_j = \varphi' + (j-3)\psi$, we obtain

$$\int c|f|^2 e^{-\varphi'}d\lambda \leqq 2\|T^*f\|_{\varphi_1}^2 + \|Sf\|_{\varphi_3}^2$$

when $f \in D_{(p,q+1)}(\Omega)$, hence when $f \in D_{T^*} \cap D_S$. Since $2\varphi_2 - \varphi - \varphi' = \varphi' - 2\psi - \varphi \geqq 0$, Cauchy-Schwarz' inequality gives if the right hand side of (4.4.2) is equal to 1

$$|(g,f)_{\varphi_2}|^2 \leqq 2^{-1}\int c|f|^2 e^{-\varphi'}d\lambda \leqq \|T^*f\|_{\varphi_1}^2 + 2^{-1}\|Sf\|_{\varphi_3}^2,\ f \in D_{T^*} \cap D_S.$$

We claim that

$$(4.4.3) \qquad |(g,f)_{\varphi_2}| \leqq \|T^*f\|_{\varphi_1},\ f \in D_{T^*}.$$

If $Sf = 0$ this is already proved. On the other hand, if f is orthogonal to the kernel of S, then f is orthogonal to the range of T so $T^*f = 0$. Since $\bar\partial g = 0$ and

$$\int |g|^2 e^{-\varphi_2}d\lambda \leqq \int |g|^2 e^{-\varphi}d\lambda < \infty$$

because $\varphi_2 - \varphi = \varphi' - \psi - \varphi \geqq \psi \geqq 0$, we have $(g,f)_{\varphi_2} = 0$ for such f which completes the proof of (4.4.3).

If we apply the Hahn-Banach theorem to the anti-linear form

$$T^*f \to (g,f)_{\varphi_2},\ f \in D_{T^*},$$

if follows that there is an element $u_a \in L^2_{(p,q)}(\Omega, \varphi_1)$ such that

$$(4.4.4) \qquad \int |u_a|^2 e^{-\varphi_1}d\lambda \leqq 1$$

and $(g,f)_{\varphi_2} = (u_a, T^*f)_{\varphi_1}$ when $f \in D_{T^*}$, hence $Tu_a = g$. We can now take a sequence $a_j \to \infty$ such that u_{a_j} is weakly L^2 convergent on Ω_a for every a to a limit u. Since $\bar\partial u_a = g$ we obtain $\bar\partial u = g$, and from (4.4.4) it follows that

$$\int_{\Omega_a} |u|^2 e^{-\varphi}d\lambda \leqq 1$$

for every a. This proves (4.4.2).

The following theorem is a simple consequence of Lemma 4.4.1 but it is often much more useful since the conditions on smoothness and strict plurisubharmonicity on φ are eliminated.

Theorem 4.4.2. *Let Ω be a pseudoconvex open set in $\mathbf{C}^n$ and φ any plurisubharmonic function in Ω. For every $g \in L^2_{(p,q+1)}(\Omega, \varphi)$ with $\bar\partial g = 0$ there is a solution $u \in L^2_{(p,q)}(\Omega, \text{loc})$ of the equation $\bar\partial u = g$ such that*

$$\int_\Omega |u|^2 e^{-\varphi}(1+|z|^2)^{-2} d\lambda \leqq \int_\Omega |g|^2 e^{-\varphi} d\lambda. \tag{4.4.5}$$

Proof. First assume that $\varphi \in C^2$. We shall apply Lemma 4.4.1 with φ replaced by $\varphi + 2\log(1+|z|^2)$, using the fact that

$$\sum_{j,k=1}^n w_j \bar w_k \frac{\partial^2}{\partial z_j \partial \bar z_k} \log(1+|z|^2) = (1+|z|^2)^{-2}(|w|^2(1+|z|^2) - |\langle w, z\rangle|^2) \geqq (1+|z|^2)^{-2}|w|^2, \tag{4.4.6}$$

which implies that we can take $c = 2(1+|z|^2)^{-2}$. Thus (4.4.2) gives (4.4.5).

In the general case we first choose a plurisubharmonic function s in Ω such that

$$\Omega_a = \{z \in \Omega;\, s(z) < a\} \Subset \Omega$$

for every $a \in \mathbf{R}$. The open sets Ω_a are also pseudoconvex. Regularizing φ as indicated in Theorem 2.6.3 we obtain C^∞ plurisubharmonic functions φ_ε defined in $\Omega_{a(\varepsilon)}$ where $a(\varepsilon) \to \infty$ as $\varepsilon \to 0$, and $\varphi_\varepsilon \downarrow \varphi$ as $\varepsilon \downarrow 0$. By the part of the theorem already proved we can now find a form $u_\varepsilon \in L^2_{(p,q)}(\Omega_{a(\varepsilon)}, \varphi_\varepsilon)$ so that $\bar\partial u_\varepsilon = g$ in $\Omega_{a(\varepsilon)}$ and

$$\int_{\Omega_{a(\varepsilon)}} |u_\varepsilon|^2 e^{-\varphi_\varepsilon}(1+|z|^2)^{-2} d\lambda \leqq \int_{\Omega_{a(\varepsilon)}} |g|^2 e^{-\varphi_\varepsilon} d\lambda \leqq \int_\Omega |g|^2 e^{-\varphi} d\lambda.$$

Since φ_ε decreases with ε this shows that the L^2 norm of u_ε over Ω_a is bounded for every fixed a. We can therefore choose a sequence $\varepsilon_j \to 0$ such that u_{ε_j} converges weakly in Ω_a for every a to a limit u in $L^2_{(p,q)}(\Omega, \text{loc})$. For every $\varepsilon > 0$ and $a \in \mathbf{R}$ we obtain

$$\int_{\Omega_a} |u|^2 e^{-\varphi_\varepsilon} d\lambda \leqq \int_\Omega |g|^2 e^{-\varphi} d\lambda$$

which implies (4.4.5), and since $\bar\partial u = g$ this completes the proof.

As a first application of Theorem 4.4.2 with $p = q = 0$ we study the extension of an analytic function defined on a linear subspace. The existence of the extension is of course a trivial matter if no growth conditions are imposed on the extended function. However, as we shall see in section 4.5, it is important to give good bounds for the extension.

Theorem 4.4.3. *Let φ be a plurisubharmonic function in $\mathbf{C}^n$ such that for some constant C*

$$|\varphi(z)-\varphi(z')| < C \quad \textit{if} \quad |z-z'| < 1. \tag{4.4.7}$$

Let Σ be a complex linear subspace of $\mathbf{C}^n$ of codimension k. For every analytic function u in Σ such that

$$\int_\Sigma |u|^2 e^{-\varphi} d\sigma < \infty, \tag{4.4.8}$$

where $d\sigma$ denotes the Lebesgue measure in Σ, there exists an analytic function U in $\mathbf{C}^n$ such that $U = u$ in Σ and

$$\int |U|^2 e^{-\varphi}(1+|z|^2)^{-3k} d\lambda \leqq 9^k \pi^k e^{kC} \int_\Sigma |u|^2 e^{-\varphi} d\sigma. \tag{4.4.9}$$

Proof. Since $\log(1+|z|^2)$ is plurisubharmonic by (4.4.6), it is enough to prove the theorem when Σ is a hyperplane and then iterate this special result k times. We may assume that Σ is the hyperplane $z_n = 0$. Then u is an analytic function of $z' = (z_1, \cdots, z_{n-1})$ and we may regard u as an analytic function in $\mathbf{C}^n$ which is independent of z_n. By (4.4.7) we have

$$\int_{|z_n|<1} |u|^2 e^{-\varphi} d\lambda \leqq \pi e^C \int_\Sigma |u|^2 e^{-\varphi} d\sigma. \tag{4.4.10}$$

Let ψ be the continuous function in $\mathbf{C}$ which is equal to 1 in the disc with radius $\frac{1}{2}$, vanishes outside the unit disc, and is a linear function of $|z|$ in between. Then $|\partial\psi/\partial\bar{z}| \leqq 1$. Writing

$$U(z) = \psi(z_n)u(z') - z_n v(z)$$

we have $U(z) = u(z')$ when $z_n = 0$ so it only remains to choose v so that v has a suitable bound and $\bar\partial U = 0$, that is,

$$\bar\partial v = z_n^{-1} u(z') \bar\partial \psi(z_n) = z_n^{-1} u(z') \partial\psi/\partial\bar{z}_n d\bar{z}_n = f. \tag{4.4.11}$$

It is clear that $\bar\partial f = 0$ and from (4.4.10) we obtain, since $\partial\psi/\partial\bar z_n = 0$ when $|z_n| < \frac{1}{2}$,

$$\int |f|^2 e^{-\varphi} d\lambda \leqq 4\pi e^C \int_\Sigma |u|^2 e^{-\varphi} d\sigma.$$

Theorem 4.4.2 now provides a solution of (4.4.11) such that

$$\int |v|^2 e^{-\varphi}(1+|z|^2)^{-2} d\lambda \leqq 4\pi e^C \int_\Sigma |u|^2 e^{-\varphi} d\sigma.$$

Combining this estimate with (4.4.10), we obtain (4.4.9).

A similar argument is used to prove the following existence theorem.

Theorem 4.4.4. *Let φ be a plurisubharmonic function in the pseudoconvex open set $\Omega \subset \mathbf{C}^n$. If $z^0 \in \Omega$ and $e^{-\varphi}$ is integrable in a neighborhood of z^0 one can find an analytic function u in Ω such that $u(z^0) = 1$ and*

$$\int_\Omega |u(z)|^2 e^{-\varphi}(1+|z|^2)^{-3n} d\lambda(z) < \infty. \tag{4.4.12}$$

Proof. We may assume that $z^0 = 0$. Choose a polydisc

$$D = \{z;\, |z_j| < r,\, j = 1, \cdots, n\} \subset \Omega$$

where $e^{-\varphi}$ is integrable, and define for $k = 0, 1, \cdots, n$

$$\Omega_k = \{z \in \Omega;\, |z_j| < r \text{ for } j > k\}.$$

We shall prove inductively that for every k there is an analytic function u_k in Ω_k with $u_k(z^0) = 1$ and

$$\int_{\Omega k} |u_k|^2 e^{-\varphi}(1+|z|^2)^{-3k} d\lambda(z) < \infty. \tag{4.4.13}$$

When $k = 0$ we can take $u_0 = 1$, and u_n will have the desired properties.

Assume that $0 < k \leqq n$ and that u_{k-1} has already been constructed. Choose $\psi \in C_0^\infty(\mathbf{C})$ so that $\psi(z) = 0$ when $|z| > r/2$ and $\psi(z) = 1$ when $|z| < r/3$, and set

$$u_k(z) = \psi(z_k) u_{k-1}(z) - z_k v(z)$$

where $\psi(z_k)u_{k-1}(z)$ is defined as 0 in $\Omega_k \backslash \Omega_{k-1}$. To make u_k analytic we must choose v as a solution of the equation

$$\bar{\partial} v = z_k^{-1} u_{k-1} \bar{\partial} \psi = f.$$

By the inductive hypothesis

$$\int_{\Omega_k} |f|^2 e^{-\varphi}(1+|z|^2)^{-3(k-1)} d\lambda < \infty$$

so it follows from Theorem 4.4.2 that v can be found so that

$$\int_{\Omega_k} |v|^2 e^{-\varphi}(1+|z|^2)^{1-3k} d\lambda < \infty.$$

Together with the inductive hypothesis on u_{k-1} this implies (4.4.13). Since $v \in C^\infty$ in a neighborhood of 0 we have $u_k(0) = u_{k-1}(0) = 1$ so u_k has the required properties.

For the application of Theorem 4.4.4 the following elementary result is required.

Theorem 4.4.5. *There is a constant C such that for every plurisubharmonic function ψ in the unit ball in $\mathbf{C}^n$ with $\psi(0) = 0$ and $\psi(z) < 1$ when $|z| < 1$*

$$\int_{|z|<\frac{1}{2}} e^{-\psi(z)} d\lambda \leqq C. \tag{4.4.14}$$

Proof. First assume that $n = 1$. We can then use the Riesz representation (Radó [1, Chap. VI])

$$2\pi\psi(z) = \int_{|\zeta|<1} \log|z-\zeta|/|1-z\bar{\zeta}|\, d\mu(\zeta) + \int_{|\zeta|=1} (1-|z|^2)|z-\zeta|^{-2} d\sigma(\zeta)$$

where $d\mu = \Delta\psi \geqq 0$ and $d\sigma \leqq |d\zeta|$ is the weak limit of ψ on the unit circle. Putting $z = 0$ we obtain $0 = \int \log|\zeta| d\mu(\zeta) + \int d\sigma(\zeta)$ or

$$\int_{|\zeta|<1} \log\frac{1}{|\zeta|}\, d\mu(\zeta) + \int_{|\zeta|=1} (|d\zeta| - d\sigma(\zeta)) = 2\pi.$$

Since both terms are non-negative it follows that

$$\int_{|\zeta|<1} \log \frac{1}{|\zeta|}\, d\mu(\zeta) \leqq 2\pi, \int |d\sigma(\zeta)| \leqq 4\pi. \tag{4.4.15}$$

Hence

$$\left| (2\pi)^{-1} \int (1-|z|^2)|z-\zeta|^{-2} d\sigma(\zeta) \right| \leqq 6 \quad \text{when} \quad |z| < \tfrac{1}{2},$$

$$a = \int_{|\zeta|<R} d\mu(\zeta)/2\pi \leqq 1/(\log(1/R)) < 2 \quad \text{if} \quad R < e^{-\frac{1}{2}}.$$

Since $e < 4$ we can choose R so that $\frac{1}{2} < R < e^{-\frac{1}{2}}$. Then we have

$$\left| \int_{|\zeta|>R} \log |z-\zeta|/|1-z\bar{\zeta}|\, d\mu(\zeta) \right| < C, \; |z| < \tfrac{1}{2},$$

for some C, and the inequality between geometric and arithmetic means shows that

$$\begin{aligned} &\exp\left(-(2\pi)^{-1} \int_{|\zeta|<R} \log |z-\zeta|/|1-z\bar{\zeta}|\, d\mu(\zeta)\right) \\ &\quad = \exp\left(\int_{|\zeta|<R} -a \log |z-\zeta|/|1-z\bar{\zeta}|\, d\mu(\zeta)/2\pi a\right) \\ &\quad \leqq \int_{|\zeta|<R} (|z-\zeta|/|1-z\bar{\zeta}|)^{-a} d\mu(\zeta)/2\pi a < C \quad \text{when} \quad |z| < \tfrac{1}{2} \end{aligned}$$

since $a < 2$. Summing up we have proved (4.4.14) when $n = 1$. The general statement follows since introducing polar coordinates we have

$$\int_{|z|<\frac{1}{2}} e^{-\psi(z)} d\lambda(z) = \int_{|\zeta|=1} dS(\zeta) \int_{|w|<\frac{1}{2}} |w|^{2n-2} e^{-\psi(w\zeta)} d\lambda(w)/2\pi$$

where $dS(\zeta)$ denotes the surface area on the unit sphere. The proof is complete.

Corollary 4.4.6. *If φ is a plurisubharmonic function in a connected pseudoconvex open set Ω and φ is not identically $-\infty$, then $e^{-\varphi}$ is locally integrable in a dense open subset G containing all points in Ω where $\varphi(z) > -\infty$. The complement of G is the set of zeros common to all analytic functions in Ω satisfying* (4.4.12).

When $n = 1$ it is easy to see from the proof of Theorem 4.4.5 that $e^{-\varphi}$ is locally integrable in the complement of the discrete set of points where $\Delta\varphi$ has a mass $\geqq 4\pi$.

In the following application of Theorem 4.4.2 we shall rely on the fact that (4.4.5) is uniform with respect to φ. By A we denote the space of entire functions in $\mathbf{C}^n$ with the topology of uniform convergence on all compact sets.

Theorem 4.4.7. *Let φ be a plurisubharmonic function in $\mathbf{C}^n$ and denote by A_φ the set of entire functions u such that for some N*

$$\int |u(z)|^2 e^{-\varphi(z)}(1+|z|^2)^{-N} d\lambda < \infty. \tag{4.4.15}$$

Then the closure of A_φ in A contains all $u \in A$ such that $|u|^2 e^{-\varphi}$ is locally integrable and it is equal to A if and only if $e^{-\varphi}$ is locally integrable.

Proof. Given an entire function U such that $|U|^2 e^{-\varphi}$ is locally integrable we shall approximate U in $\{z; |z| < R\}$ by functions in A_φ. To do so we choose $\chi \in C_0^\infty(\mathbf{C}^n)$ so that $\chi(z) = 1$ when $|z| < R + 2$ and set $V = \chi U$. Then

$$f = \bar\partial V = U\bar\partial\chi = 0 \quad \text{when} \quad |z| < R+2,$$

and f has compact support. To make the norm of f small we set for $t > 0$

$$\varphi_t(z) = \varphi(z) + \max(0, t\log(|z|/(R+1))).$$

Then it is clear that φ_t is plurisubharmonic and that

$$\int |f|^2 \exp(-\varphi_t) d\lambda \to 0, \; t \to \infty,$$

so it follows from Theorem 4.4.2 that we can find u_t with $\bar\partial u_t = f$ and

$$\int |u_t|^2 \exp(-\varphi_t)(1+|z|^2)^{-2} d\lambda \to 0 \quad \text{as} \quad t \to \infty.$$

In particular, $\bar\partial u_t = 0$ when $|z| < R + 1$, and

$$\int_{|z|<R+1} |u_t|^2 d\lambda \to 0$$

so $u_t(z) \to 0$ uniformly when $t \to \infty$ and $|z| < R$. Since

$$\int |u_t|^2(1+|z|^2)^{-2-(t/2)}e^{-\varphi(z)}d\lambda(z) < \infty,$$

the analytic function $V - u_t$ satisfies (4.4.15) with $N = 2 + (t/2)$. Hence $V - u_t \in A_\varphi$ and converges uniformly to U when $|z| < R$ and $t \to \infty$. This proves the theorem, for every function in A_φ must vanish at z if $e^{-\varphi}$ is not integrable in any neighborhood of z.

Note that the proof of Theorem 4.4.7 gives an alternative proof of Lemma 4.3.1.

4.5. Analytic functionals. Already in the proof of Theorem 4.4.4 we considered linear forms on the space of entire analytic functions but we shall make a more systematic study in this section.

Definition 4.5.1. *Elements in the dual space A' of the space $A = A(\mathbf{C}^n)$ of entire functions, equipped with the topology of uniform convergence on compact sets, are called analytic functionals. An analytic functional μ is said to be carried by a compact set K if for every neighborhood ω of K there is a constant C_ω such that*

$$|\mu(f)| \le C_\omega \sup_\omega |f|, \qquad f \in A. \tag{4.5.1}$$

From the definition of the topology in A it follows that for every $\mu \in A'$ there is some compact ω such that (4.5.1) holds. Hence μ has some compact carrier, but we shall give an example below where there is no carrier which is contained in all others. This is in contrast with the notion of support of a function or distribution.

Linear combinations of the exponential functions $\exp\langle z, \zeta\rangle$, where $\zeta \in \mathbf{C}^n$ and $\langle z, \zeta\rangle = z_1\zeta_1 + \cdots + z_n\zeta_n$, are dense in A, for if we differentiate $e^{\langle z, \zeta\rangle}$ with respect to ζ and put $\zeta = 0$ afterward, we conclude that z^α is in the closure of the linear hull of the exponentials for every α. An analytic functional μ is therefore uniquely determined by its values for the exponentials, that is, by the Laplace transform:

Definition 4.5.2. *If $\mu \in A'$, we define the Laplace transform by*

$$\tilde{\mu}(\zeta) = \mu_z(e^{\langle z, \zeta\rangle}), \qquad \zeta \in \mathbf{C}^n. \tag{4.5.2}$$

The Laplace transform is an entire analytic function of ζ, for the continuity properties of μ show that we can differentiate under the μ-sign to prove that $\tilde{\mu} \in C^\infty$ and that $\partial\tilde{\mu}/\partial\bar{\zeta}_j = 0$ for all j. From (4.5.1) we obtain the estimate

$$|\tilde{\mu}(\zeta)| \leq C_\omega \exp(\sup_{z\in\omega} \operatorname{Re}\langle z, \zeta\rangle).$$

Set

$$H_K(\zeta) = \sup_{z\in K} \operatorname{Re}\langle z, \zeta\rangle. \tag{4.5.3}$$

Then H_K is a convex positively homogeneous function of ζ, and if K is convex we have

$$K = \{z\,;\operatorname{Re}\langle z, \zeta\rangle \leq H_K(\zeta), \zeta \in \mathbf{C}^n\}. \tag{4.5.4}$$

(This well-known property of support functions of convex sets is proved by applying the Hahn–Banach theorem to the convex set $\{(t,\zeta) \in \mathbf{R}\times\mathbf{C}^n;$ $H(\zeta) \leq t\}$.) The remarks we have now made prove the first half of the following

Theorem 4.5.3. *If $\mu \in A'$ and μ is carried by the compact set K, then $M(\zeta) = \tilde{\mu}(\zeta)$ is an entire analytic function and for every $\delta > 0$ there is a constant C_δ such that*

$$|M(\zeta)| \leq C_\delta \exp(H_K(\zeta) + \delta|\zeta|), \qquad \zeta \in \mathbf{C}^n. \tag{4.5.5}$$

Conversely, if K is a convex compact set and M an entire function satisfying (4.5.5) *for every $\delta > 0$, there exists an analytic functional μ carried by K such that $\tilde{\mu} = M$.*

In particular, the Laplace transformation is a one-to-one mapping of A' onto the space of all entire functions M which are of exponential type, that is, which for suitable constants a and C have the bound

$$|M(\zeta)| \leq C \exp a|\zeta|.$$

We shall first give the classical proof of Theorem 4.5.3 when $n = 1$ and then a proof depending on Theorem 4.4.3 which is valid for every n.

Proof of Theorem 4.5.3 when $n = 1$. Let M be a function satisfying (4.5.5) and form the power series expansion

$$M(\zeta) = \sum_0^\infty a_k \zeta^k/k!.$$

Choose A and C so that $|M(\zeta)| \leq C\, e^{A|\zeta|}$. Then the Cauchy inequalities give

$$|a_k|/k! \leq C\, e^{AR}/R^k$$

for every $R > 0$. Choosing $R = k/A$ we minimize the right-hand side and obtain from Stirling's formula with a constant C_1

$$|a_k| \le C(Ae)^k k!/k^k \le C\,C_1{}^k. \tag{4.5.6}$$

Now we can write

$$\zeta^k/k! = (2\pi i)^{-1}\int_\gamma e^{z\zeta} z^{-k-1}\,dz,$$

where γ is a convex curve with the origin in its interior. Hence we have formally

$$M(\zeta) = \sum_0^\infty a_k\zeta^k/k! = (2\pi i)^{-1}\int_\gamma e^{z\zeta}B(z)\,dz, \tag{4.5.7}$$

where

$$B(z) = \sum_0^\infty a_k z^{-k-1} \tag{4.5.8}$$

is called the Borel transform of M. Now it follows from (4.5.6) that this series is uniformly convergent when $|z| > C_1 + 1$, so (4.5.7) is in fact valid if γ lies in this set.

We shall prove that B can be continued to an analytic function in $\complement K$. Admitting this fact for a moment we set

$$\mu(f) = (2\pi i)^{-1}\int_\gamma f(z)\,B(z)\,dz,$$

where γ is an arbitrary convex curve surrounding K. This defines an analytic functional which does not depend on the choice of γ, and μ is therefore carried by K. By (4.5.7) we have $\tilde{\mu} = M$.

To construct the analytic continuation we form with arbitrary fixed $\zeta \in \mathbf{C}$

$$B_\zeta(z) = \int_0^\infty M(t\zeta)\,e^{-tz\zeta}\zeta\,dt.$$

This integral is absolutely convergent and defines an analytic function in the half plane

$$\{z\,;H(\zeta) < \operatorname{Re}\zeta z\}, \tag{4.5.9}$$

and when $C_1|\zeta| < \operatorname{Re}\zeta z$ it follows from (4.5.6) that we can compute $B_\zeta(z)$

by using the power series expansion for M, which gives that $B_\zeta(z) = B(z)$ then. Hence any two functions B_{ζ_1} and B_{ζ_2} coincide in some open subset of the intersection of the half planes where they are defined and they must therefore agree in their common domain of definition. It follows that the functions B_ζ define an analytic continuation of B to the set of all z such that $H(\zeta) < \operatorname{Re} \zeta z$ for some ζ, which by (4.5.4) is the complement of K.

Note that the proof also gives a one-to-one correspondence between analytic functionals and functions which are analytic in a neighborhood of infinity and vanish there.

General proof of Theorem 4.5.3. Let K_ε be the set of points at euclidean distance $< \varepsilon$ from K. We wish to construct a continuous function ψ vanishing outside K_ε such that the analytic functional defined by the measure $\psi\, d\lambda$ has the Laplace transform M, that is,

$$M(\zeta) = \int \psi(z)\, e^{\langle z,\zeta \rangle}\, d\lambda(z). \tag{4.5.10}$$

Let $\hat{\psi}$ be the Fourier-Laplace transform of ψ considered as a function of $2n$ real variables

$$\hat{\psi}(\theta_1, \cdots, \theta_{2n}) = \int \psi(x)\, e^{-i(x_1\theta_1 + \ldots + x_{2n}\theta_{2n})}\, dx_1 \cdots dx_{2n}.$$

This is an analytic function of $2n$ complex variables. The condition (4.5.10) means that, if we set $\hat{\psi} = \Psi$,

$$\Psi(i\zeta_1, -\zeta_1, i\zeta_2, -\zeta_2, \cdots, i\zeta_n, -\zeta_n) = M(\zeta_1, \cdots, \zeta_n), \tag{4.5.11}$$

so the construction of ψ is an extension problem of the type considered in Theorem 4.4.3. If ψ has its support in K_ε we must have

$$|\hat{\psi}(\theta)| \le C \exp \varphi_\varepsilon(\theta), \tag{4.5.12}$$

where

$$\varphi_\varepsilon(\theta) = \sup_{x \in K_\varepsilon} (x_1 \operatorname{Im} \theta_1 + \cdots + x_{2n} \operatorname{Im} \theta_{2n}). \tag{4.5.13}$$

Conversely, if Ψ is an entire analytic function in $\mathbf{C}^{2n}$ such that

$$|\Psi(\theta)| \le C(1 + |\theta|)^{-2n-1} \exp \varphi_\varepsilon(\theta), \qquad \theta \in \mathbf{C}^{2n}, \tag{4.5.12$'$}$$

it follows from the Paley–Wiener theorem in its most elementary form that Ψ is the Fourier–Laplace transform of a continuous function ψ vanishing outside K_ε, and if Ψ satisfies (4.5.11) the theorem will be proved. (The proof of the Paley–Wiener theorem consists simply in forming the inverse Fourier transform

$$\psi(x) = (2\pi)^{-2n} \int_{\mathbf{R}^{2n}} e^{i\langle x, \theta' + i\theta'' \rangle} \Psi(\theta' + i\theta'')\, d\theta'.$$

A change of integration contour shows that this is independent of θ''. If we replace θ'' by $t\theta''$ and let $t \to +\infty$, it follows that $\psi(x) = 0$ when $\langle x, \theta'' \rangle > \varphi_\varepsilon(\theta'')$, and from an analogue of (4.5.4) we conclude that ψ vanishes outside K_ε. By Fourier's inversion formula the Laplace transform of the function ψ is equal to Ψ.)

To construct Ψ we denote by u the function

$$(i\zeta_1, -\zeta_1, \cdots, i\zeta_n, -\zeta_n) \to M(\zeta_1, \cdots, \zeta_n),$$

which is an analytic function in an n-dimensional subspace Σ of $\mathbf{C}^{2n}$. If $\theta = (i\zeta_1, -\zeta_1, \cdots, i\zeta_n, -\zeta_n)$, we obtain from (4.5.13)

$$\varphi_\varepsilon(\theta) = \sup_{x \in K_\varepsilon}(x_1 \operatorname{Re} \zeta_1 - x_2 \operatorname{Im} \zeta_1 + \cdots) = \sup_{z \in K_\varepsilon} \operatorname{Re}\langle z, \zeta \rangle = H_K(\zeta) + \varepsilon|\zeta|.$$

Hence (4.5.5) implies that

$$|u(\theta)| \leq C_\delta \exp(\varphi_\varepsilon(\theta) + (\delta - \varepsilon)|\zeta|), \qquad \delta > 0, \theta \in \Sigma,$$

and choosing $\delta < \varepsilon$, we obtain

$$\int_\Sigma |u|^2 e^{-2\varphi_\varepsilon} d\sigma < \infty.$$

Since φ_ε is convex and therefore plurisubharmonic, and since φ_ε is uniformly continuous, it follows from Theorem 4.4.3 that there is an entire analytic function U in $\mathbf{C}^{2n}$ such that $U = u$ in Σ and

$$\int |U(\theta)|^2 e^{-2\varphi_\varepsilon(\theta)} (1 + |\theta|^2)^{-3n} d\lambda < \infty,$$

where $d\lambda$ is now the Lebesgue measure in $\mathbf{C}^{2n}$. By Theorem 2.2.3 this implies that

$$|U(\theta)| \leq C\, e^{\varphi_\varepsilon(\theta)} (1 + |\theta|)^{3n}. \tag{4.5.14}$$

Now set

$$\Psi(\theta) = U(\theta)\hat{\chi}(\theta)$$

where $\hat{\chi}$ is the Fourier–Laplace transform of a non-negative function $\chi \in C_0^\infty(\mathbf{C}^n)$ with support in $\{z; |z| < \varepsilon\}$, such that χ only depends on $|z|$ and the integral of χ equals 1. This means that $\hat{\chi}(0) = 1$ and that $\hat{\chi}(\theta)$ is a function of $\theta_1{}^2 + \cdots + \theta_{2n}{}^2$, which implies that $\hat{\chi} = 1$ in Σ. Hence Ψ satisfies (4.5.11), and since by the Paley–Wiener theorem

$$|\hat{\chi}(\theta)| \leq C_N(1 + |\theta|)^{-N}\, e^{\varepsilon|\mathrm{Im}\,\theta|}$$

for every N (this is proved by partial integration), the estimate (4.5.12)′ with ε replaced by 2ε follows from (4.5.14).

Example. The function $M(\zeta) = \cos(2\sqrt{\zeta_1\zeta_2})$ is entire and of exponential type. Since $|\cos t| \leq e^{|t|}$ for every complex number t, it follows that

$$|M(\zeta)| \leq \exp(a_1|\zeta_1| + a_2|\zeta_2|) \quad \text{if } a_1a_2 = 1 \text{ and } a_1, a_2 > 0.$$

Hence the analytic functional μ for which $\tilde{\mu} = M$ is carried by the set $\{(z_1,z_2); |z_1| \leq a_1, |z_2| \leq a_2\}$ if $a_1a_2 = 1$. The intersection of these carriers is the origin, but it is clear that μ is not carried by the origin since $M(\zeta_1,\zeta_1)$ is of exponential type 2. Hence there need not exist a smallest convex set which carries M except in the case $n = 1$ where the Borel transform of M can be continued analytically to the complement of the intersection of all convex sets carrying μ, which proves that μ is carried by the intersection.

Notes. Existence theorems for the Cauchy–Riemann equations in domains of holomorphy (stated as solution of the first Cousin problem) were first proved by Oka [2]. He also proved an approximation theorem for functions analytic in a neighborhood of a holomorph-convex compact subset. For a smaller class of domains this is due to Weil [3]. The identity of pseudoconvex domains and domains of holomorphy was proved much later by Oka [5], Bremermann [2] and Norguet [1]. It is an important feature of the methods used here that they solve the first Cousin problem directly in pseudoconvex domains. This makes it easy to prove that these are domains of holomorphy. In doing so we also prove that they are characterized by the existence of solutions to the equation $\bar{\partial}u = f$ for every form with $\bar{\partial}f = 0$ (cf. Serre [3]). Techniques similar to those used here were first proposed by Garabedian and Spencer [1] in analogy with the Hodge–de Rham–Kodaira decomposition of forms on Riemannian manifolds. The basic a priori estimates were first proved by Morrey [1] for (0, 1) forms and by Kohn [1] in the general case. (A simplification due to Ash [1] of the proof of Kohn [1] will be used in Chapter V.) Kohn [2] also proved some theorems on boundary regularity required in his approach and that of Morrey; a simplified proof has been given by Kohn and Nirenberg [1]. The technique of using weight functions to modify the L^2 norms goes back to Carleman in other areas of the theory of partial differential equations. It was introduced in the present context by Hörmander [1] to avoid the boundary difficulties referred to above and to prove sharper results. Related arguments are due to Andreotti and Vesentini [1]. More precise existence theorems in L^2 norms are given in section 4.4 with a few applications. Theorem 4.4.4 is due to Bombieri [1]; for further applications we refer to Skoda [1, 2]. The important feature of Theorem 4.4.2, the main result, is its uniformity with respect to the weight functions and pseudo-convex sets considered. For an application of the latter property we refer to Hörmander-Wermer [1]. However, in this context one obtains better results from the existence theory for the equation $\bar{\partial}u = f$ in maximum norms recently developed by Henkin, Ramirez, Grauert-Lieb, Kerzman, Øvrelid and others. (See Harvey-Wells [1] and the references given there.) These techniques fall outside the scope of this book. As a simple application of the results of section 4.4 we prove in section 4.5 a theorem on analytic functionals due to Pólya [1] for one complex variable and to Ehrenpreis [2] and Martineau [1] in general. Using results of Martineau [1], Kiselman [1] has studied when there is a unique minimal carrier for an analytic functional. For applications of analytic functionals, see e.g. Ehrenpreis [3], Malgrange [3].

Chapter V

STEIN MANIFOLDS

Summary. Until this chapter we have studied function theory in open subsets of $\mathbf{C}^n$ only. However, this is too special for many purposes. For example, the simultaneous analytic continuation of all functions analytic in an open subset of $\mathbf{C}^n$ may lead to functions defined in a Riemann domain which is spread over $\mathbf{C}^n$ with several sheets just as a Riemann surface may be in the case $n = 1$. We therefore introduce a class of complex analytic manifolds, the Stein manifolds, whose definition is modeled on the properties of domains of holomorphy in $\mathbf{C}^n$. The definitions are given in section 5.1, and in section 5.2 we show that the existence and approximation theorems for the Cauchy–Riemann equations given in sections 4.2 and 4.3 can be extended to this more general case. In section 5.3 we prove that a Stein manifold can be represented concretely as a closed submanifold of a space $\mathbf{C}^N$ of sufficiently high dimension. Conversely, such manifolds are always Stein manifolds. The maximal simultaneous analytic continuation of the analytic functions in a domain in $\mathbf{C}^n$ is studied in section 5.4. We show that it leads to a Stein manifold spread over $\mathbf{C}^n$ and that such manifolds can be characterized by a pseudoconvexity condition. (Sections 5.3 and 5.4 are not necessary for reading the later parts of this book.) At last the Cousin problems are stated in section 5.5 where we solve them for Stein manifolds. In section 5.6 we extend the results of section 5.2 to sections of analytic vector bundles over Stein manifolds. This is an essential preparation for the theory of coherent analytic sheaves to be given in Chapter VII. In section 5.7 finally we show that an analytic structure can also be defined by prescribing at each point of a manifold what differentials shall be of type (0,1) in such a way that a certain integrability condition is fulfilled. The result is important in the study of perturbations of complex structures, but the reader may bypass it here without any loss of continuity.

5.1. Definitions. A Hausdorff topological space Ω is called a manifold of dimension n if every point in Ω has a neighborhood which is homeomorphic to an open set in $\mathbf{R}^n$. The concept of complex analytic manifolds is defined by means of a family of such homeomorphisms:

Definition 5.1.1. *A manifold Ω (of dimension $2n$) is called a complex analytic manifold of complex dimension n if there is given a family $\mathscr{F}$ of homeomorphisms κ, called complex analytic coordinate systems, of open sets $\Omega_\kappa \subset \Omega$ on open sets $\tilde{\Omega}_\kappa \subset \mathbf{C}^n$ such that*

(i) *If κ and $\kappa' \in \mathscr{F}$, then the mapping*

$$\kappa'\kappa^{-1} : \kappa(\Omega_\kappa \cap \Omega_{\kappa'}) \to \kappa'(\Omega_\kappa \cap \Omega_{\kappa'})$$

between open sets in $\mathbf{C}^n$ is analytic. (Interchanging κ and κ' we find that the inverse mapping is also analytic.)

(ii)
$$\bigcup_{\kappa \in \mathscr{F}} \Omega_\kappa = \Omega.$$

(iii) *If κ_0 is a homeomorphism of an open set $\Omega_0 \subset \Omega$ on an open set in $\mathbf{C}^n$ and the mapping*

$$\kappa\kappa_0^{-1} : \kappa_0(\Omega_0 \cap \Omega_\kappa) \to \kappa(\Omega_0 \cap \Omega_\kappa)$$

as well as its inverse are analytic for every $\kappa \in \mathscr{F}$, it follows that $\kappa_0 \in \mathscr{F}$.

The condition (iii) is in a way superfluous. For if $\mathscr{F}$ satisfies (i) and (ii), we can extend $\mathscr{F}$ in one and only one way to a family $\mathscr{F}'$ satisfying (i), (ii), (iii). In fact, the only such family $\mathscr{F}'$ is the set of all mappings satisfying the condition (iii) relative to $\mathscr{F}$. A complex analytic structure can thus be defined by an arbitrary family $\mathscr{F}$ satisfying (i) and (ii), but if condition (iii) is dropped, there are many families defining the same structure. Such a family is called a complete set of complex analytic coordinate systems, and two such sets are called equivalent if they define the same structure.

We shall say that n complex valued functions $(z_1, \cdots, z_n)$ defined in a neighborhood of a point w in Ω are a local coordinate system at w if they define a mapping of a neighborhood of w into $\mathbf{C}^n$ which is a coordinate system in the sense defined above. If $f_1, \cdots, f_n$ are analytic functions in a neighborhood of $z(w) = (z_1(w), \cdots, z_n(w))$ in $\mathbf{C}^n$, then $(f_1(z), \cdots, f_n(z))$ is another system of coordinates at w if and only if $\det(\partial f_i/\partial z_j)_{i,j=1}^n \neq 0$ at $z(w)$. This follows from Theorem 2.1.2.

Definition 5.1.2. *Let Ω_1 and Ω_2 be complex analytic manifolds. Then a mapping $f: \Omega_1 \to \Omega_2$ is called analytic if $\kappa_2 \circ f \circ \kappa_1^{-1}$ is analytic (where it is defined) for all coordinate systems κ_1 in Ω_1 and κ_2 in Ω_2.*

It is of course sufficient to choose only coordinate systems in complete sets of coordinate systems in Ω_1 and Ω_2. In particular, we have now defined the concept of analytic functions in a complex analytic manifold Ω; the set of such functions with the topology of uniform convergence on compact subsets of Ω will be denoted by $A(\Omega)$. It is obvious that Ω is a Fréchet space if Ω is countable at infinity, that is, if there exists a countable number of compact subsets $K_1, K_2, \cdots$ such that every compact subset of Ω is contained in some K_j. In fact, the topology in $A(\Omega)$ is then defined by the seminorms

$$A(\Omega) \ni f \to \sup_{K_j} |f|, \qquad j = 1, 2, \cdots$$

and the completeness is obvious.

It is clear that every open subset of a complex manifold Ω has a structure of complex analytic manifold, so the concept of an analytic function (mapping) on an open subset is also well defined. Note that if f is analytic in $\tilde{\Omega}_\kappa \subset \mathbf{C}^n$, then $f \circ \kappa$ is analytic in Ω_κ. Hence by the definition of a complex analytic manifold, analytic functions do exist locally. We shall now define a class of manifolds where there is a good supply of globally defined analytic functions. As we shall see, complex function theory in such manifolds behaves essentially as in domains of holomorphy in $\mathbf{C}^n$.

Definition 5.1.3. *A complex analytic manifold Ω of dimension n which is countable at infinity is said to be a Stein manifold if*

(α) *Ω is holomorph-convex, that is,*

$$\hat{K} = \{z; z \in \Omega, |f(z)| \le \sup_K |f| \text{ for every } f \in A(\Omega)\}$$

is a compact subset of Ω for every compact subset K of Ω.

(β) *If z_1 and z_2 are different points in Ω, then $f(z_1) \neq f(z_2)$ for some $f \in A(\Omega)$.*

(γ) *For every $z \in \Omega$, one can find n functions $f_1, \cdots, f_n \in A(\Omega)$ which form a coordinate system at z.*

Example. By Theorem 2.5.5, every domain of holomorphy in $\mathbf{C}^n$ is a Stein manifold.

To give another example, we need a definition.

Definition 5.1.4. *A subset V of a complex analytic manifold Ω of dimension n is called an analytic submanifold of dimension m if*

(i) *V is closed.*

(ii) *In a neighborhood ω of an arbitrary point $v \in V$ there exist local coordinates $z_1, \cdots, z_n$ such that $\omega \cap V = \{w; w \in \omega, z_{m+1}(w) = \cdots = z_n(w) = 0\}$.*

We can define a natural analytic structure on V by means of the coordinate systems $(z_1, \cdots, z_m)$ when $(z_1, \cdots, z_n)$ is a coordinate system for Ω with the stated property. If $f_1, \cdots, f_n$ is an arbitrary coordinate system for Ω at $v \in V$, then one can always find m among these functions which form a coordinate system for V at v. In fact, since the Jacobian

$$\det(\partial f_i/\partial z_j) \neq 0 \qquad (i, j = 1, \cdots, n)$$

at $z(v)$, one can choose $i_1, \cdots, i_m$ so that

$$\det(\partial f_{i_\mu}/\partial z_\nu) \neq 0 \qquad (\mu, \nu = 1, \cdots, m).$$

Hence the restrictions of $f_{i_1}, \cdots, f_{i_m}$ to V form a local coordinate system at v.

Theorem 5.1.5. *Every submanifold of a Stein manifold is a Stein manifold.*

Proof. (α), (β) are trivial since the restriction to a submanifold of a function which is analytic in the whole manifold is necessarily analytic. (γ) follows from the remark just made.

This theorem would have been false for every smaller class of manifolds containing all spaces $\mathbf{C}^n$. In fact, we shall prove in section 5.3 that every Stein manifold of dimension n can be embedded as a submanifold of $\mathbf{C}^{2n+1}$.

Finally, we prove that Theorem 2.6.11 can be extended to Stein manifolds.

Theorem 5.1.6. *Let Ω be a Stein manifold, K a compact subset of Ω, and ω an open neighborhood of $\hat{K}$. Then there exists a function $\varphi \in C^\infty(\Omega)$ such that*

(a) *φ is strictly plurisubharmonic.*
(b) *$\varphi < 0$ in K but $\varphi > 0$ in $\complement\omega$.*
(c) *$\{z; z \in \Omega,\ \varphi(z) < c\} \subset\subset \Omega$ for every $c \in \mathbf{R}$.*

Note that the notion of strict plurisubharmonicity is well defined for functions on a complex analytic manifold since it is invariant under analytic changes of variables. (See the proof of Theorem 2.6.4.)

Proof of Theorem 5.1.6. By condition (α) in Definition 5.1.3 we can choose a sequence $K_1 = \hat{K}, K_2, \cdots$ of compact subsets of Ω such that $\hat{K}_j = K_j$, $\cup K_j = \Omega$, and K_j is in the interior of K_{j+1} for every j. Let ω_j be an open set with $K_j \subset \omega_j \subset K_{j+1}$ and $\omega_1 \subset \omega$. Since $\hat{K}_j = K_j$, we can for every j choose functions $f_{jk} \in A(\Omega)$, $k = 1, \cdots, k_j$ with absolute value < 1 in K_j so that $\max_k |f_{jk}(z)| > 1$, $z \in K_{j+2} \setminus \omega_j$. By raising f_{jk} to high powers we may even arrange that

$$\sum_{k=1}^{k_j} |f_{jk}(z)|^2 < 2^{-j}, \qquad z \in K_j, \tag{5.1.1}$$

$$\sum_{k=1}^{k_j} |f_{jk}(z)|^2 > j, \qquad z \in K_{j+2} \setminus \omega_j. \tag{5.1.2}$$

In view of condition (γ) in Definition 5.1.3, we may also assume that among the functions f_{jk}, $k = 1, \cdots, k_j$, it is possible to find n functions forming a system of local coordinates at any point in K_j. Now form

$$\varphi(z) = \sum_{j,k} |f_{jk}(z)|^2 - 1.$$

The sum converges by (5.1.1), and $\varphi(z) > j - 1$ when $z \in \complement\, \omega_j$ by (5.1.2). φ is in fact in $C^\infty(\Omega)$, for the series

$$\sum_{j,k} f_{jk}(z)\overline{f_{jk}(\zeta)}$$

converges uniformly on compact subsets of $\Omega \times \Omega$, so the sum is analytic in z and its complex conjugate is analytic in ζ. It is also clear that φ is plurisubharmonic, and φ is strictly plurisubharmonic since, if for some z

$$\sum_{l=1}^{n} w_l\, \partial f_{jk}(z)/\partial z_l = 0 \quad \text{for all } j \text{ and } k,$$

then $w = 0$ because there exist n functions f_{jk} forming a local system of coordinates at z. This completes the proof.

Note that even $\log(2 + \varphi)$ is strictly plurisubharmonic by Corollary 1.6.8 and Theorem 1.6.12.

Remark. In proving Theorem 5.1.6 we only used conditions (α) and (γ) in Definition 5.1.3. We shall see in section 5.2 that this leads to the conclusion that condition (β) is a consequence of the other conditions.

5.2. L^2 estimates and existence theorems for the $\bar{\partial}$ operator. Let Ω be a complex analytic manifold of complex dimension n, which is countable at infinity. The decomposition of differential forms into forms of type (p,q) and the definition of the $\bar{\partial}$ operator can immediately be extended to forms and functions on the manifold Ω, for all these concepts are invariant for analytic changes of coordinates (see section 2.1).

In order to extend the Hilbert space techniques used in Chapter IV, we must introduce hermitian norms on differential forms in Ω. We therefore choose a hermitian metric on Ω, that is, a Riemannian metric which in any analytic coordinate system with coordinates $z_1, \cdots, z_n$ has the form

$$\sum_{j,k=1}^{n} h_{jk}\, dz_j\, d\bar{z}_k,$$

where h_{jk} is a positive definite hermitian matrix with C^∞ coefficients. The existence of such a hermitian structure is trivial locally, and is immediately proved in the large by means of a partition of unity. The invariant element of volume defined by the structure we denote by dV. (For definitions, see also Weil [1].)

If f is a form of type (1,0) and $f = \Sigma_1^n f_j\, dz_j$ in a local coordinate system, we set

$$\langle f,f \rangle = \sum h^{kj} f_j \bar{f}_k,$$

where (h^{jk}) is the inverse of (h_{jk}). This hermitian form is invariantly defined, for

$$\langle f,f \rangle = \sup \left|\sum f_j\, dz_j\right|^2 / \sum h_{jk}\, dz_j\, d\bar{z}_k.$$

By the Gram-Schmidt orthogonalization process every point in Ω has a neighborhood U where there are n forms $\omega^1, \cdots, \omega^n$ of type (1,0) with C^∞ coefficients such that at every point in U

$$\langle \omega^j, \omega^k \rangle = \delta_{jk}, \qquad j, k = 1, \cdots, n.$$

If we set $f = \Sigma f_j \omega^j$, it follows that $\langle f,f \rangle = \Sigma_1^n |f_j|^2$. More generally, a differential form f of type (p,q) can be written in a unique way as a sum

$$f = \sum_{|I|=p}' \sum_{|J|=q}' f_{I,J}\, \omega^I \wedge \bar{\omega}^J,$$

where $f_{I,J}$ are antisymmetric both in I and in J and Σ' means that summation is extended only over increasing multi-indices. We can define $\langle f,f \rangle$ by

$$\langle f,f \rangle = |f|^2 = \sum{}' |f_{I,J}|^2 = \frac{1}{p!q!} \sum |f_{I,J}|^2,$$

for this definition is independent of the choice of orthonormal basis $\omega^1, \cdots, \omega^n$. (We leave the verification of this fact as an exercise.)

As in Lemma 4.1.3, we choose once and for all a sequence $\eta_1, \eta_2, \cdots$ of functions in $C_0^\infty(\Omega)$ such that $0 \le \eta_\nu \le 1$ and $\eta_\nu = 1$ on any compact subset of Ω when v is large. As a substitute for condition (4.1.6) we shall modify the hermitian metric so that

$$|\bar\partial\eta_v| \leq 1 \quad \text{on } \Omega, \qquad v = 1, 2, \cdots. \tag{5.2.1}$$

To see that this is possible we only have to note that given a hermitian metric we can choose a positive C^∞ function M on Ω such that

$$|\bar\partial\eta_v| \leq M \quad \text{for all } v.$$

In fact, this means only a finite number of lower bounds for M on any compact subset of Ω. If we replace the metric by

$$M^2 \sum h_{jk}\, dz_j\, d\bar z_k$$

the condition (5.2.1) will be fulfilled when the norm of η_v is defined with respect to the new metric. *From now on we keep the hermitian structure and the sequence η_v fixed.*

Let φ be a function in $C^2(\Omega)$, and let $L^2_{(p,q)}(\Omega,\varphi)$ be the space of all (equivalence classes of) forms of type (p,q), such that the coefficients are measurable in any local coordinate system, and

$$\|f\|_\varphi{}^2 = \int |f|^2 e^{-\varphi}\, dV < \infty.$$

The operator $\bar\partial$ defines linear, closed, densely defined operators

$$T: L^2_{(p,q)}(\Omega,\varphi) \to L^2_{(p,q+1)}(\Omega,\varphi)$$

and

$$S: L^2_{(p,q+1)}(\Omega,\varphi) \to L^2_{(p,q+2)}(\Omega,\varphi).$$

Lemma 5.2.1. *$D_{(p,q+1)}(\Omega)$ is dense in $D_{T^*} \cap D_S$ for the graph norm*

$$f \to \|f\|_\varphi + \|T^*f\|_\varphi + \|Sf\|_\varphi.$$

Proof. We shall essentially repeat the proof of Lemma 4.1.3. Since

$$S(\eta_v f) - \eta_v Sf = \bar\partial\eta_v \wedge f, \qquad f \in D_S,$$

it follows from (5.2.1) that

$$|S(\eta_v f) - \eta_v Sf|^2 \leq |f|^2, \qquad f \in D_S.$$

Further, we have

$$((T^*(\eta_v f) - \eta_v T^*f),u)_\varphi = -(f,\bar\partial\eta_v \wedge u)_\varphi \quad \text{if } f \in D_{T^*} \text{ and } u \in D_{(p,q)},$$

which gives

$$|T^*(\eta_\nu f) - \eta_\nu T^*f|^2 \le |f|^2.$$

Hence $\eta_\nu f \to f$ in the graph norm if $f \in D_{T^*} \cap D_S$. It is therefore sufficient to approximate elements in $D_{T^*} \cap D_S$ which have compact support, and by means of a partition of unity we reduce the proof to the case when the support lies in a coordinate patch. In that case we can use the following classical lemma of Friedrichs.

Lemma 5.2.2. *Let $\varphi \in C_0^\infty(\mathbf{R}^N)$ and $\int \varphi\, dx = 1$, let $v \in L^2(\mathbf{R}^N)$ have compact support and let a be a C^1 function in a neighborhood of the support of v. Put*

$$(J_\varepsilon v)(x) = \int v(x - \varepsilon y)\varphi(y)\, dy.$$

Then $aD_k J_\varepsilon v - J_\varepsilon(aD_k v) \to 0$ in L^2 when $\varepsilon \to 0$.

Here $aD_k v$ is defined in the sense of distribution theory, $D_k = \partial/\partial x_k$.

Proof. The statement is obvious if $v \in C^1$. It is therefore enough to prove the inequality

$$\|aD_k J_\varepsilon v - J_\varepsilon(aD_k v)\|_{L^2} \le C\|v\|_{L^2},$$

for small ε, assuming that a has bounded derivatives in the whole of $\mathbf{R}^N$. A simple computation gives

$$W_\varepsilon(x) = aD_k J_\varepsilon v(x) - J_\varepsilon(aD_k v)(x) = \int v(x - \varepsilon y)((a(x) - a(x - \varepsilon y))\varphi_k(y)/\varepsilon + (D_k a)(x - \varepsilon y))\varphi(y)\, dy.$$

Here we have written $\varphi_k = D_k\varphi$. If C is a bound for $|\text{grad}\, a|$, it follows that

$$|W_\varepsilon(x)| \le C \int |v(x - \varepsilon y)|(|y||\varphi_k(y)| + |\varphi(y)|)\, dy,$$

so Minkowski's inequality for integrals gives

$$\|W_\varepsilon\|_{L^2} \le C \int (|y||\varphi_k(y)| + |\varphi(y)|)\, dy \|v\|_{L^2}.$$

End of proof of Lemma 5.2.1. From Lemma 5.2.2 it follows immediately that if in a coordinate patch where f has compact support we apply the operator J_ε to (each component of) f, then

$$\|T^*(J_\varepsilon f) - J_\varepsilon T^* f\|_\varphi \to 0, \qquad \varepsilon \to 0,$$

hence in view of Lemma 4.1.4

$$\|T^*(J_\varepsilon f) - T^* f\|_\varphi \to 0 \quad \text{when } \varepsilon \to 0, \text{ if } f \in D_{T^*}.$$

Similarly we prove that $S(J_\varepsilon f) \to Sf$ if $f \in D_S$. (This is simpler since S has constant coefficients if the coordinates are analytic.) The proof of Lemma 5.2.1 is thus completed.

Let U be a coordinate patch in Ω where there is a C^∞ orthonormal basis $\omega^1, \cdots, \omega^n$ for forms of type (1,0). In terms of this basis we shall now give expressions for the operators S and T^* acting on forms $f \in D_{(p,q+1)}$ with support in U. This will prove that Lemma 4.2.1 remains essentially unchanged for such forms.

If $u \in C^1(U)$, we can write

$$du = \sum_1^n \partial u/\partial\omega^i \omega^i + \sum_1^n \partial u/\partial\bar\omega^i \bar\omega^i$$

as a definition of the first-order linear differential operators $\partial/\partial\omega^i$ and $\partial/\partial\bar\omega^i$. Then we have $\bar\partial u = \Sigma_1^n \partial u/\partial\bar\omega^i\bar\omega^i$, and if $f = \Sigma' f_{I,J}\omega^I \wedge \bar\omega^J$ it follows that

$$\bar\partial f = \sum_{I,J}{}' \sum_i \partial f_{I,J}/\partial\bar\omega^i\bar\omega^i \wedge \omega^I \wedge \bar\omega^J + \cdots,$$

where the dots indicate terms in which no $f_{I,J}$ is differentiated; they occur because $\bar\partial\omega^i$ and $\bar\partial\bar\omega^j$ need not be 0. If the sum is denoted by Af, we obviously have $|\bar\partial f - Af| \le C|f|$ where C is independent of f when the support of f lies in a fixed compact subset of U.

Now let $u \in D_{(p,q)}$ have its support in U and form

$$(5.2.2) \qquad \int \langle T^*f, u\rangle e^{-\varphi}\, dV = \int \langle f, \bar\partial u\rangle e^{-\varphi}\, dV$$
$$= (-1)^p \sum_{I,K}{}' \sum_j \int f_{I,jK}\overline{\partial u_{I,K}/\partial\bar\omega^j}\, e^{-\varphi}\, dV + \cdots,$$

where dots again indicate terms involving no differentiations. We shall integrate by parts here. First note that, with the notation

$$\delta_j w = e^{\varphi}\, \partial(we^{-\varphi})/\partial\omega^j,$$

Green's formula gives

$$\int \partial v/\partial\bar\omega^j \bar w\, e^{-\varphi}\, dV = -\int v\overline{\delta_j w}\, e^{-\varphi}\, dV + \int \sigma_j v\, \bar w\, e^{-\varphi}\, dV, \quad v, w \in C_0^\infty(U)$$

where $\sigma_j \in C^\infty(U)$. Integrating by parts in (5.2.2), we obtain

$$(5.2.3) \quad T^*f = (-1)^{p-1} \sum_{I,K}{}' \sum_{j=1}^n \delta_j f_{I,jK}\omega^I \wedge \bar\omega^K + \cdots = Bf + \cdots,$$

where the dots indicate terms where no $f_{I,jK}$ is differentiated and which do not involve φ. Hence $|T^*f - Bf| \le C|f|$ when f has its support in a fixed compact subset of U. With another constant independent of f and of φ, we thus obtain

$$\|Af\|_\varphi^2 + \|Bf\|_\varphi^2 \le 2(\|Sf\|_\varphi^2 + \|T^*f\|_\varphi)^2 + C\|f\|_\varphi^2. \tag{5.2.4}$$

The arguments which led to (4.2.3) still apply, so we get

$$\|Af\|_\varphi^2 + \|Bf\|_\varphi^2 = \int \sum_{I,J}{}' \sum_{j=1}^{n} |\partial f_{I,J}/\partial\overline{\omega}^j|^2 e^{-\varphi}\, dV + \tag{5.2.5}$$
$$\int \sum_{I,K}{}' \sum_{j,k=1}^{n} (\delta_j f_{I,jK}\overline{\delta_k f_{I,kK}} - \partial f_{I,jK}/\partial\overline{\omega}^k \overline{\partial f_{I,kK}/\partial\overline{\omega}^j})\, e^{-\varphi}\, dV.$$

Before repeating the integration by parts next performed in section 4.2, we must consider the commutators of the operators $\partial/\partial\overline{\omega}^j$ and δ_k to obtain a substitute for (4.2.6). Thus, let w be a smooth function in U and form

$$\bar\partial\partial w = \bar\partial \sum_{k=1}^{n} \partial w/\partial\omega^k \omega^k = \sum_{j,k=1}^{n} \partial^2 w/\partial\overline{\omega}^j\partial\omega^k \overline{\omega}^j \wedge \omega^k + \sum_{i=1}^{n} \partial w/\partial\omega^i \bar\partial\omega^i.$$

Since $\bar\partial\omega^i$ is a form of type (1,1), we may write

$$\bar\partial\omega^i = \sum_{j,k=1}^{n} c^i_{jk}\overline{\omega}^j \wedge \omega^k, \tag{5.2.6}$$

which gives

$$\bar\partial\partial w = \sum_{j,k} (\partial^2 w/\partial\overline{\omega}^j\partial\omega^k + \sum_i c^i_{jk}\, \partial w/\partial\omega^i)\overline{\omega}^j \wedge \omega^k.$$

If we replace w by $\overline{w}$ and take complex conjugates of all terms, we also obtain

$$\partial\bar\partial w = \sum_{j,k} (\partial^2 w/\partial\omega^j\partial\overline{\omega}^k + \sum_i \bar c^i_{jk}\, \partial w/\partial\overline{\omega}^i)\omega^j \wedge \overline{\omega}^k.$$

The identity $\bar\partial\partial w = -\partial\bar\partial w$ therefore implies that

$$w_{kj} = \partial^2 w/\partial\overline{\omega}^j\partial\omega^k + \sum_i c^i_{jk}\, \partial w/\partial\omega^i = \partial^2 w/\partial\omega^k\partial\overline{\omega}^j + \sum_i \bar c^i_{kj}\, \partial w/\partial\overline{\omega}^i, \tag{5.2.7}$$

where the left-hand side is a definition. Note that with this notation we have

$$\partial\bar\partial w = \sum w_{jk}\omega^j \wedge \overline{\omega}^k.$$

Hence w is plurisubharmonic precisely when the form $\Sigma\, w_{jk} f_j \bar f_k$ is positive definite.

From (5.2.7) it follows that

$$(\delta_k \partial w/\partial \bar\omega^j - \partial \delta_k w/\partial \bar\omega^j) = \partial^2 \varphi/\partial \bar\omega^j \partial \omega^k w + \sum_i c^i_{jk}\, \partial w/\partial \omega^i - \sum_i \bar c^i_{kj}\, \partial w/\partial \bar\omega^i,$$

or if we use the definition of δ_i and (5.2.7) again, with w replaced by φ,

$$(\delta_k \partial w/\partial \bar\omega^j - \partial \delta_k w/\partial \bar\omega^j) = \varphi_{kj} w + \sum_i c^i_{jk} \delta_i w - \sum_i \bar c^i_{kj}\, \partial w/\partial \bar\omega^i. \tag{5.2.8}$$

Using Green's formula and (5.2.8), we now integrate by parts in (5.2.5). This gives, in view of (5.2.4),

$$\sum_{I,K}{}' \int \sum_{j,k=1}^{n} f_{I,jK} \overline{f_{I,kK}} \varphi_{jk}\, e^{-\varphi}\, dV + \sum_{I,J}{}' \int \sum_{j=1}^{n} |\partial f_{I,J}/\partial \bar\omega^j|^2\, e^{-\varphi}\, dV \tag{5.2.9}$$
$$\le 2(\|Sf\|^2_\varphi + \|T^* f\|^2_\varphi) + C\|f\|^2_\varphi + t_1 + t_2 + t_3,$$

where

$$t_1 = \sum_{I,K}{}' \sum_{i,j,k=1}^{n} \int f_{I,jK} \overline{\bar c^i_{jk} \delta_i f_{I,kK}}\, e^{-\varphi}\, dV,$$

$$t_2 = -\sum_{I,K}{}' \sum_{i,j,k=1}^{n} \int f_{I,jK} c^i_{kj} \overline{\partial f_{I,kK}/\partial \bar\omega^i}\, e^{-\varphi}\, dV,$$

$$t_3 = \sum_{I,K}{}' \sum_{j,k} \int (f_{I,jK} \bar\sigma_j \overline{\delta_k f_{I,kK}} - f_{I,jK} \sigma_k \overline{\partial f_{I,kK}/\partial \bar\omega^j})\, e^{-\varphi}\, dV.$$

In t_1 and in those terms in t_3 which involve the operators δ_i, another integration by parts leads to terms involving only the differential operators $\partial/\partial \bar\omega^i$. If these are estimated by Cauchy–Schwarz' inequality, it follows from (5.2.9) that

$$\int \sum_{I,J}{}' |f_{I,J}|^2 \lambda\, e^{-\varphi}\, dV + \tfrac{1}{2} \sum_{I,J}{}' \sum_{j=1}^{n} \int |\partial f_{I,J}/\partial \bar\omega^j|^2\, e^{-\varphi}\, dV \tag{5.2.10}$$
$$\le 2(\|T^* f\|^2_\varphi + \|Sf\|^2_\varphi + C\|f\|^2_\varphi)$$

for all $f \in D_{(p,q+1)}$ with support in a fixed compact subset of U. Here C is a constant and λ denotes the smallest eigenvalue of the hermitian symmetric form

$$\sum \varphi_{jk} t_j \bar t_k. \tag{5.2.11}$$

(Note that λ is independent of the choice of basis $\omega^1, \cdots, \omega^n$, for a change of basis means only a unitary transformation of the variables $t_1, \cdots, t_n$.) The estimate (5.2.10) is of course mainly of interest when φ is plurisubharmonic.

We can now easily give a global version of (5.2.10).

Theorem 5.2.3. *There exists a continuous function C on Ω such that*

$$\int (\lambda - C)|f|^2 e^{-\varphi} \, dV \leq 4(\|T^*f\|_\varphi^2 + \|Sf\|_\varphi^2), \qquad f \in D_{(p,q+1)}(\Omega). \tag{5.2.12}$$

Here φ is an arbitrary function in $C^2(\Omega)$ and λ is the lowest eigenvalue of the form (5.2.11), *which is also a continuous function in* Ω.

The constant 4 here could be replaced by any number larger than 1.

Proof of Theorem 5.2.3. Let U_j, $j = 1, 2, \cdots$ be coordinate patches in Ω where (5.2.10) is applicable, chosen so that they form a locally finite covering of Ω. (That is, $\cup U_j = \Omega$ and no compact set in Ω meets more than a finite number of sets U_j.) Choose $\psi_j \in C_0^\infty(U_j)$ so that

$$\sum \psi_j^2 = 1 \text{ in } \Omega.$$

(If $\psi_j' \in C_0^\infty(U_j)$ and $\psi^2 = \Sigma \psi_j'^2 > 0$ everywhere in Ω, we can take $\psi_j = \psi_j'/\psi$.) Now apply (5.2.10) to $\psi_j f$. This gives

$$\int \psi_j^2 |f|^2 \lambda e^{-\varphi} \, dV \leq 4(\|\psi_j T^*f\|_\varphi^2 + \|\psi_j Sf\|_\varphi^2) + C_j \int_{U_j} |f|^2 e^{-\varphi} \, dV.$$

Adding these estimates, we immediately obtain (5.2.12).

Theorem 5.2.3 gives a perfect substitute for Lemma 4.2.1. Our discussion can therefore proceed along the lines of sections 4.2 and 4.3, so we shall make it brief.

Theorem 5.2.4. *Let Ω be a complex manifold where there exists a strictly plurisubharmonic function φ such that $\{z; z \in \Omega, \varphi(z) < c\} \subset\subset \Omega$ for every $c \in \mathbf{R}$. Then the equation $\bar\partial u = f$ has (in the weak sense) a solution $u \in L^2_{(p,q)}(\Omega,\text{loc})$ for every $f \in L^2_{(p,q+1)}(\Omega,\text{loc})$ such that $\bar\partial f = 0$.*

Proof. Let us replace the function φ in (5.2.12) by $\chi(\varphi)$ where χ is a convex increasing function. The lower bound λ for the form (5.2.11) can then be replaced by $\chi'(\varphi)\lambda$ (see the proof of Theorem 4.2.2), so we obtain from Theorem 5.2.3

$$\int (\chi'(\varphi)\lambda - C)|f|^2 e^{-\chi(\varphi)} \, dV \leq 4(\|T^*f\|_{\chi(\varphi)}^2 + \|Sf\|_{\chi(\varphi)}^2), \qquad f \in D_{(p,q+1)}(\Omega).$$

If χ' is chosen so rapidly increasing that

$$\chi'(\varphi)\lambda - C \geq 4, \tag{5.2.13}$$

it follows that

$$\|f\|_{\chi(\varphi)}^2 \leq \|T^*f\|_{\chi(\varphi)}^2 + \|Sf\|_{\chi(\varphi)}^2, \qquad f \in D_{T^*} \cap D_S, \tag{5.2.14}$$

if we apply Lemma 5.2.1. (The operator T^* is of course the adjoint of T with respect to the norms $\| \ \|_{\chi(\varphi)}$.) If χ satisfies (5.2.13), Lemma 4.1.1 (see also (4.1.5)) now shows that the equation $\bar\partial u = f$ has a solution $u \in L^2_{(p,q)}(\Omega,\chi(\varphi))$ for every $f \in L^2_{(p,q+1)}(\Omega,\chi(\varphi))$ satisfying the equation $\bar\partial f = 0$, and u can be chosen so that

$$\|u\|_{\chi(\varphi)} \leq \|f\|_{\chi(\varphi)}. \tag{5.2.15}$$

This proves the theorem.

The spaces $W^s_{(p,q)}, 0 \leq s \leq \infty$, introduced after Theorem 4.2.2 are obviously invariant under analytic changes of coordinates, so if we have a manifold Ω, the space $W^s_{(p,q)}(\Omega,\text{loc})$ can be defined as the set of forms belonging to $W^s_{(p,q)}$ in every coordinate patch. The proof of Theorem 4.2.5 applies with evident modifications since, for forms $f \in D_{T^*} \cap D_S$ with support in a coordinate patch, we obtain from (5.2.10) estimates of the derivatives $\partial f_{I,J}/\partial\bar\omega^j$ and therefore of the derivatives $\partial f_{I,J}/\partial\bar z_j$ if z_j are the local coordinates. Thus we leave as an exercise for the reader to supply the details of the proof of the following theorem:

Theorem 5.2.5. *Let Ω be a complex manifold where there exists a strictly plurisubharmonic function $\varphi \in C^\infty(\Omega)$ such that $\{z; z \in \Omega, \varphi(z) < c\} \subset\subset \Omega$ for every c. Then the equation $\bar\partial u = f$ has a solution $u \in W^{s+1}_{(p,q)}(\Omega,\text{loc})$ for every $f \in W^s_{(p,q+1)}(\Omega,\text{loc})$ such that $\bar\partial f = 0$. Every solution of the equation $\bar\partial u = f$ has this property when $q = 0$.*

The last statement of course contains nothing new beyond Theorem 4.2.5.

Corollary 5.2.6. *Under the hypotheses of Theorem 5.2.5, the equation $\bar\partial u = f$ has a solution $u \in C^\infty_{(p,q)}(\Omega)$ for every $f \in C^\infty_{(p,q+1)}(\Omega)$ such that $\bar\partial f = 0$.*

Since Theorem 2.7.10 can obviously be extended to manifolds, we obtain in view of Theorem 5.1.6

Theorem 5.2.7. *If Ω is a Stein manifold of dimension n, then $H^r(\Omega,\mathbf{C}) = 0$ when $r > n$.*

We also obtain approximation theorems:

Theorem 5.2.8. *Let Ω be a complex manifold and φ a strictly plurisubharmonic function in Ω such that $K_c = \{z; z \in \Omega, \varphi(z) \leq c\} \subset\subset \Omega$ for every real number c. Every function which is analytic in a neighborhood of K_0 can then be approximated uniformly on K_0 by functions in $A(\Omega)$.*

Proof. We can apply the proof of Lemma 4.3.1, using the estimate

(5.2.14) instead of (4.2.10). (The function φ in Theorem 5.2.8 is the function p in Lemma 4.3.1.) The details will not be repeated.

Combining Theorem 5.2.8 with Theorem 5.1.6, we obtain

Corollary 5.2.9. *If Ω is a Stein manifold and K a compact subset of Ω with $\hat{K} = K$, then every function which is analytic in a neighborhood of K can be approximated uniformly on K by functions in $A(\Omega)$.*

It is now possible for us to prove a converse of Theorem 5.1.6.

Theorem 5.2.10. *A complex manifold Ω is a Stein manifold if and only if there exists a strictly plurisubharmonic function $\varphi \in C^\infty(\Omega)$ such that $\Omega_c = \{z ; z \in \Omega, \varphi(z) < c\} \subset\subset \Omega$ for every real number c. The sets $\bar{\Omega}_c$ are then $A(\Omega)$ convex.*

Proof. Theorem 5.1.6 states that functions φ with these properties exist in every Stein manifold. Conversely, suppose that such a function φ exists in the manifold Ω. We have to prove the three conditions (α), (β), (γ) in Definition 5.1.3. First we give a lemma.

Lemma 5.2.11. *For every $z^0 \in \Omega$ there is a neighborhood ω_0 and an analytic function $u_0 \in A(\omega_0)$ such that $u_0(z^0) = 0$ and*

$$\operatorname{Re} u_0(z) < \varphi(z) - \varphi(z^0) \quad \text{if } z^0 \neq z \in \omega_0.$$

Proof. Let $z_1, \cdots, z_n$ be local coordinates at z^0, such that the coordinates of z^0 are all 0. The Taylor expansion of φ can be written

$$\varphi(z) = \varphi(0) + \operatorname{Re} u_0(z) + \sum_{j,k=1}^{n} \partial^2\varphi(0)/\partial z_j \partial \bar{z}_k \, z_j \bar{z}_k + O(|z|^3),$$

where u_0 is a polynomial of degree ≤ 2 with $u_0(0) = 0$. Since the hermitian form is positive definite, it follows that

$$\varphi(z) > \varphi(0) + \operatorname{Re} u_0(z), \qquad z \neq 0,$$

in a neighborhood of the origin. This proves the lemma.

End of proof of Theorem 5.2.10. First note that the hypotheses in the theorem are also satisfied by Ω_c, with φ replaced by $1/(c - \varphi)$. We therefore have an existence theorem for the $\bar{\partial}$ operator (Corollary 5.2.6) in Ω_c for every c.

Let z^0 be an arbitrary point in Ω. We shall prove that there exist local coordinates at z^0 consisting of functions in $A(\Omega)$, that z^0 does not belong to the $A(\Omega)$-hull of $\bar{\Omega}_c$ for any $c < \varphi(z^0)$, and that for any other point z^1 with $\varphi(z^1) \leq \varphi(z^0)$ there is a function $f \in A(\Omega)$ with $f(z^0) \neq f(z^1)$. This will of course prove the theorem.

Choose u_0 and ω_0 according to Lemma 5.2.11 so that $z^1 \notin \omega_0$, and ω_0 is covered by one set of local coordinates. Then take other neighborhoods ω_1 and ω_2 of z^0 so that

$$\omega_1 \subset\subset \omega_2 \subset\subset \omega_0.$$

Let ψ be a function in $C_0^\infty(\omega_2)$ which is equal to 1 in ω_1. We shall use ψ as a standard cutoff function. Since the support of $\bar{\partial}\psi$ lies in $\bar{\omega}_2 \setminus \omega_1$, we can choose $a > \varphi(z^0)$ and $\varepsilon > 0$ so that

$$\text{(5.2.16)} \qquad \operatorname{Re} u_0(z) < -\varepsilon \quad \text{if } z \in \operatorname{supp} \bar{\partial}\psi \text{ and } \varphi(z) < a.$$

From the proof of Theorem 5.2.4 for the $\bar{\partial}$ operator in Ω_a, it follows that there is a function $\varphi_a \in C^\infty(\Omega_a)$, bounded from below in Ω_a, such that the equation $\bar{\partial}v = f$ for every $f \in L^2_{(0,1)}(\Omega_a, \varphi_a)$ with $\bar{\partial}f = 0$ has a solution $v \in L^2(\Omega_a, \varphi_a)$ with

$$\text{(5.2.17)} \qquad \|v\|_{\varphi_a} \leq \|f\|_{\varphi_a}.$$

By Corollary 5.2.6, v is infinitely differentiable if f is infinitely differentiable.

Now let u be an analytic function in ω_0 and set with a large positive parameter t

$$\text{(5.2.18)} \qquad u_t = \psi u\, e^{tu_0} - v_t.$$

We want to choose v_t so that $u_t \in A(\Omega_a)$ and v_t is small. To do so we note that u_t is analytic if

$$\text{(5.2.19)} \qquad \bar{\partial}v_t = u\, e^{tu_0}\bar{\partial}\psi = f_t$$

where the last equality is a definition. By (5.2.16) we have the estimate $\|f_t\|_{\varphi_a} = O(e^{-\varepsilon t})$, for fixed u, so application of (5.2.17) shows that (5.2.19) has a solution v_t with

$$\text{(5.2.20)} \qquad \|v_t\|_{\varphi_a} = O(e^{-\varepsilon t}), \qquad t \to +\infty.$$

The equation (5.2.19) means in particular that v_t is analytic in the complement of the support of $\bar{\partial}\psi$ in Ω_a. By Theorem 2.2.3 we therefore obtain $u_t(z^1) = v_t(z^1) \to 0$ and $u_t(z^0) = u(z^0) + v_t(z^0) \to u(z^0)$ when $t \to \infty$. Since $u(z^0)$ can be chosen equal to 1, it follows that $u_t(z^1) \neq u_t(z^0)$ if t is large. By Theorem 5.2.8 we can approximate u_t on $\Omega_{\varphi(z^0)}$ so closely by a function $U \in A(\Omega)$ that $U(z^0) \neq U(z^1)$. This proves condition (β). Furthermore, (5.2.20) implies that

$$\int_{\Omega_c} |u_t|^2\, dV \to 0, \qquad t \to \infty,$$

for every $c < \varphi(z^0)$. Hence Theorem 2.2.3 shows that $u_t \to 0$ uniformly on compact subsets of Ω_c. If $c' < c$, it follows that $|u_t| < \frac{1}{2}$ on $\Omega_{c'}$ for large t while $u_t(z^0) \to 1$. Approximating u_t by functions in $A(\Omega)$, using Theorem 5.2.8, we conclude that z^0 is not in the $A(\Omega)$-hull of $\bar{\Omega}_{c'}$ for any $c' < \varphi(z^0)$. Hence the $A(\Omega)$-hull of $\bar{\Omega}_{c'}$ is equal to $\bar{\Omega}_{c'}$ for every c'. This proves condition (α).

Finally note that if the function $u \in A(\omega_0)$ is chosen so that $u(z^0) = 0$, the facts that $\partial u_t = \partial u - \partial v_t$ at z^0, and that $\partial v_t \to 0$ at z^0 by Theorem 2.2.3, show that ∂u_t converges to ∂u at z^0 when $t \to \infty$. If $u^1, \cdots, u^n$ are a coordinate system at z^0, formed by functions vanishing there, the Jacobian of the corresponding functions $u_t^{\,1}, \cdots, u_t^{\,n} \in A(\Omega_a)$ with respect to $u^1, \cdots, u^n$ must therefore converge to 1 when $t \to \infty$. Using Theorem 5.2.8 we can approximate these functions so closely by functions $U^1, \cdots, U^n \in A(\Omega)$ that the Jacobian of $U^1, \cdots, U^n$ with respect to $u^1, \cdots, u^n$ is also $\neq 0$ at z^0. Hence condition (γ) in Definition 5.1.3 is also fulfilled.

Corollary 5.2.12. *The condition* (β) *in Definition* 5.1.3 *is a consequence of the other hypotheses.*

Proof. This is an immediate consequence of Theorem 5.2.10 and the remark following Theorem 5.1.6.

5.3. Embedding of Stein manifolds. Let Ω be a complex analytic manifold of dimension n. Every $f = (f_1, \cdots, f_N) \in A(\Omega)^N$ defines an analytic map

$$\Omega \ni z \to (f_1(z), \cdots, f_N(z)) \in \mathbf{C}^N. \tag{5.3.1}$$

Definition 5.3.1. *The map* (5.3.1) *is called regular if it has rank n at every point in Ω, that is, if for any point in Ω there is a coordinate system formed by n of the functions $f_1, \cdots, f_N$. If the inverse image of every compact subset of $\mathbf{C}^N$ is a compact subset of Ω, the map is called proper.*

The condition of regularity can of course also be stated as follows: If $(z_1, \cdots, z_n)$ are local coordinates, then the Jacobian matrix $(\partial f_i/\partial z_j)$, $i = 1, \cdots, N, j = 1, \cdots, n$, has rank n at every point in the coordinate patch.

It is clear that the range of a proper map (5.3.1) is closed, for every compact set is mapped on a compact set. If, in addition, the map is regular and one-to-one, the range is an analytic submanifold of $\mathbf{C}^N$ which is isomorphic to Ω. Our aim is to construct such a map when Ω is a Stein manifold.

Lemma 5.3.2. *If K is a compact subset of a complex manifold satisfying conditions* (β) *and* (γ) *in Definition* 5.1.3, *then one can for some large N find $f \in A(\Omega)^N$ so that f is regular and one-to-one on K.*

Proof. By the Borel–Lebesgue lemma and condition (γ) in Definition 5.1.3 we can choose $(f_1, \cdots, f_k) \in A(\Omega)^k$ so that n among these functions form a coordinate system at any point in K. Then there is a neighborhood V of the diagonal in $K \times K$ such that $(z', z'') \in V$ and $f_j(z') = f_j(z'')$, $j = 1, \cdots, k$, implies $z' = z''$. Using (β), we can now choose a finite number of functions $f_{k+1}, \cdots, f_N$ in $A(\Omega)$ so that $f_j(z') = f_j(z'')$, $j = k + 1, \cdots, N$, and $(z', z'') \in K \times K \setminus V$ implies $z' = z''$. Altogether we then have a map with the required properties.

Our next purpose is to show that N need not be chosen larger than $2n + 1$. This follows by an argument of Whitney.

Lemma 5.3.3. *If K is a compact subset of a complex manifold Ω of dimension n, and if $f = (f_1, \cdots, f_N) \in A(\Omega)^N$, then f maps K on a compact set of Lebesgue measure 0 in $\mathbf{C}^N$ if $N > n$.*

Proof. That f maps K on a compact set follows from the continuity of f. In proving that the range has measure 0, we may assume that K is contained in one coordinate patch, with coordinates $(z_1, \cdots, z_n)$, for K can be covered by a finite number of coordinate patches. Since $f(z + \zeta) = f(z) + O(|\zeta|)$, the measure of the set in $\mathbf{C}^N$ on which f maps a cube I in $\mathbf{C}^n$ with side ε is therefore $O(\varepsilon^{2N}) = m(I)O(\varepsilon^2)$ because $N > n$. Now we can cover K by cubes with side ε and total measure $< m(K) + 1$, so it follows that $m(f(K)) < (m(K) + 1)O(\varepsilon^2)$. Hence $f(K)$ is a null set.

Lemma 5.3.4. *If $f \in A(\Omega)^{N+1}$, $N \geq 2n$, is a regular map on the compact subset K of Ω, then one can find $(a_1, \cdots, a_N) \in \mathbf{C}^N$ arbitrarily close to the origin so that*

$$(f_1 - a_1 f_{N+1}, \cdots, f_N - a_N f_{N+1}) \in A(\Omega)^N$$

is a regular map on K. In fact, this is true for all $a \in \mathbf{C}^N$ outside a set of measure 0.

Proof. We may assume that K is contained in one coordinate patch with coordinates $(z_1, \cdots, z_n)$. The vector $a \in \mathbf{C}^N$ has to be chosen so that, if

$$\sum_1^n \lambda_k(\partial f_j/\partial z_k - a_j \partial f_{N+1}/\partial z_k) = 0, \qquad j = 1, \cdots, N,$$

at some point in K and for some $\lambda \in \mathbf{C}^N$ it follows that $\lambda = 0$. With

$a_{N+1} = 1$ and $\mu = \Sigma_1^n \lambda_k \partial f_{N+1}/\partial z_k$, this condition can be rephrased as follows: The equations

$$\sum_1^n \lambda_k \partial f_j/\partial z_k = \mu a_j, \qquad j = 1, \cdots, N+1,$$

shall imply that $\lambda = 0$. Since the matrix $(\partial f_j/\partial z_k)_{j=1,\cdots,N}^{k=1,\cdots,n}$ has rank n, it is therefore sufficient to choose a so that $(a,1)$ is not in the range of the map

$$\text{(5.3.2)} \qquad \mathbf{C}^n \times K \ni (\lambda, z) \to \left\{ \sum_1^n \lambda_k \partial f_j/\partial z_k \right\}_{j=1}^{N+1} \in \mathbf{C}^{N+1}.$$

If we first restrict λ to a ball $|\lambda| \leq v$, $v = 1, 2, \cdots$, it follows from Lemma 5.3.3 that the range of the map (5.3.2) is of measure 0, for $N + 1 > 2n$, and since its intersections with the planes $a_{N+1} =$ constant are homothetic, they must all be of $2N$-dimensional measure 0. Hence we have proved the lemma.

Lemma 5.3.5. *If* $f \in A(\Omega)^{N+1}$, $N \geq 2n + 1$, *is a regular one-to-one map on the compact subset K of* Ω, *then one can find* $(a_1, \cdots, a_N) \in \mathbf{C}^N$ *arbitrarily close to the origin so that*

$$(f_1 - a_1 f_{N+1}, \cdots, f_N - a_N f_{N+1}) \in A(\Omega)^N$$

is a regular one-to-one map on K. In fact, this is true for all $a \in \mathbf{C}^N$ *outside a set of measure* 0.

Proof. We know from Lemma 5.3.4 that the map is regular on K when a avoids a set of measure 0. We want to choose a so that for $z', z'' \in K$ the equations

$$f_j(z') - a_j f_{N+1}(z') = f_j(z'') - a_j f_{N+1}(z''), \qquad j = 1, \cdots, N,$$

imply $z' = z''$. With $a_{N+1} = 1$ and $\lambda = f_{N+1}(z') - f_{N+1}(z'')$, we can write them in the form

$$f_j(z') - f_j(z'') = \lambda a_j, \qquad j = 1, \cdots, N+1.$$

Hence it is sufficient to show that $a_1, \cdots, a_{N+1}$ can be chosen with $a_{N+1} = 1$ so that these $N + 1$ equations imply that $\lambda = 0$, and therefore that $z' = z''$ since f is one-to-one. But the range of the map

$$\mathbf{C} \times K \times K \ni (\mu, z', z'') \to \mu(f_1(z') - f_1(z''), \cdots, f_{N+1}(z') - f_{N+1}(z'')) \in \mathbf{C}^{N+1}$$

is a null set since $N + 1 > 1 + 2n$. This proves the lemma.

Remark. Note that the geometrical meaning of the proof of Lemmas 5.3.4 and 5.3.5 is that we have projected $\mathbf{C}^{N+1}$ on the subspace $\mathbf{C}^N$ defined by $z_{N+1} = 0$ along a direction which is not a tangent (respectively not even a chord) of $f(K)$.

It is now easy to prove the existence of regular one-to-one maps.

Theorem 5.3.6. *Let Ω be a complex manifold which is countable at infinity and satisfies conditions (β) and (γ) in Definition* 5.1.3. *Then*

(a) *The set of all $f \in A(\Omega)^N$ which do not give a regular map of Ω into $\mathbf{C}^N$ is of the first category if $N \geq 2n$.*

(b) *The set of all $f \in A(\Omega)^N$ which do not give a regular one-to-one map of Ω into $\mathbf{C}^N$ is of the first category if $N \geq 2n + 1$.*

We recall that a set in a complete metrizable space is of the first category if it is contained in the union of countably many closed sets with no interior. A set of the first category has no interior, that is, its complement is dense.

Proof of Theorem 5.3.6. We first prove (a). Since Ω is the union of countably many compact sets, it is sufficient to prove that for every compact subset K of Ω, the set M of all $f \in A(\Omega)^N$ which are not regular on K is of the first category. Now M is obviously closed, for if $f_j \in A(\Omega)^N$ and $f_j \to f$ when $j \to \infty$, then, if f_j is not regular at $z_j \in K$, it follows that f is not regular at any limit point of the sequence z_j. It is therefore sufficient to prove that M has no interior point. To do so we choose according to Lemma 5.3.2 functions $g_1, \cdots, g_r \in A(\Omega)$ so that $g = (g_1, \cdots, g_r)$ is a regular map into $\mathbf{C}^r$ on K. For every $f \in A(\Omega)^N$ we can now apply Lemma 5.3.4 repeatedly to the map (f,g) into $\mathbf{C}^{N+r}$ and conclude that

$$f_j' = f_j + \sum_1^r a_{jk} g_k, \qquad j = 1, \cdots, N$$

is a regular map on K for suitable, arbitrarily small coefficients a_{jk}. Thus f' is not in M, so f is not an interior point of M. This completes the proof of (a). The proof of (b) is exactly parallel except for the fact that it depends on Lemma 5.3.5 instead of Lemma 5.3.4. It may therefore be left to the reader.

It remains to discuss the existence of proper maps, which is much harder. First note that if we have a proper map f of Ω into $\mathbf{C}^N$, then $\{z; z \in \Omega, |f_j(z)| < R, j = 1, \cdots, N\}$ is relatively compact in Ω for every R and these analytic polyhedra, all defined by no more than N

inequalities, exhaust Ω. What we shall do first is therefore to discuss analytic polyhedra in a Stein manifold.

We shall call an open relatively compact set $P \subset \Omega$ an analytic polyhedron of *order* N if for some $f_j \in A(\Omega)$, $j = 1, \cdots, N$, the set P is a union of components of the open set

$$\{z; z \in \Omega, |f_j(z)| < 1, j = 1, \cdots, N\}.$$

Lemma 5.3.7. *If Ω is a Stein manifold, K a compact subset of Ω with $K = \hat{K}$, and ω a neighborhood of K, then there exists an analytic polyhedron P with $K \subset P \subset\subset \omega$.*

Proof. We may assume that ω is relatively compact in Ω. For every $z \in \partial\omega$, we can find $f \in A(\Omega)$ so that $|f| < 1$ in K but $|f(z)| > 1$. By the Borel–Lebesgue lemma we can therefore choose $f_1, \cdots, f_N \in A(\Omega)$ so that

$$\{z; z \in \Omega, |f_j(z)| < 1, j = 1, \cdots, N\}$$

contains K but does not meet $\partial\omega$. Hence the intersection of this set with ω is an analytic polyhedron P with the required properties.

The next step is to decrease the order of the polyhedron with a device due to Bishop.

Lemma 5.3.8. *Let K be a compact set and P an analytic polyhedron of order $N + 1$ in Ω such that $K \subset P$. If $N \geq 2n$, there exists an analytic polyhedron P' of order N such that $K \subset P' \subset P$.*

Proof. Let P be a union of components of the set

$$\{z; z \in \Omega, |f_j(z)| < 1, j = 1, \cdots, N + 1\}.$$

Choose numbers $c_0 < c_1 < c_2 < c_3 < 1$ so that $|f_j(z)| < c_0$ for $j = 1, \cdots, N + 1$ when $z \in K$. We can choose $f_1', \cdots, f'_{N+1}$ with $f'_{N+1} = f_{N+1}$ so that

$$(f_1'/f_{N+1}, \cdots, f_N'/f_{N+1}) \text{ is of rank } n \text{ on } \{z; z \in \bar{P}, |f_{N+1}(z)| \geq c_2\},$$

and j_j' is so close to f_j that $|f_j'(z)| < c_0$ in K for $j = 1, \cdots, N$ and

$$\omega = \{z; z \in P, |f_j'(z)| < c_3, j = 1, \cdots, N + 1\} \subset\subset P.$$

In fact, this follows immediately from the proof of Theorem 5.3.6 since $N \geq 2n$; we can choose f_j'/f_{N+1} as f_j/f_{N+1} plus a linear combination with small coefficients of suitable functions in $A(\Omega)$.

Now consider the open set

$$\Delta_\nu = \{z; z \in \Omega, |f_j'(z)^\nu - f_{N+1}(z)^\nu| < c_1{}^\nu, j = 1, \cdots, N\},$$

where v is a positive integer which will be chosen later. We shall prove that the union P_v' of the components of $\triangle_v$ which intersect K has the desired properties if v is large enough.

First note that $z \in K$ implies

$$|f_j'(z)^v - f_{N+1}(z)^v| < 2c_0{}^v < c_1{}^v$$

if v is sufficiently large. Hence $K \subset \triangle_v$ for large v. If we can prove that $P_v' \subset \omega$ when v is large, we will have an analytic polyhedron of order N with all the required properties. If P_v' is not contained in ω, then some point $z \in P_v'$ must be on the boundary of ω, for every component of P_v' intersects K and therefore contains points in ω. If $|f_{N+1}(z)| < c_2$, then

$$|f_j'(z)|^v < c_2{}^v + c_1{}^v < c_3{}^v \quad \text{when } j \leq N \text{ if } v \text{ is large,}$$

which contradicts the assumption that $z \in \partial\omega$. Hence z is in the compact set

$$L = \{z; z \in \partial\omega, |f_{N+1}(z)| \geq c_2\}.$$

Let L_1 be a compact subset of L contained in a coordinate patch with coordinates $z_1, \cdots, z_n$. If $z \in L_1 \cap \triangle_v$ we have, with $F_j = f_j'/f_{N+1}$,

$$|F_j(z)^v - 1| < (c_1/c_2)^v, \qquad j = 1, \cdots, N.$$

We shall prove that this implies that

$$\max_{j=1,\cdots,N} |f_j'(z+\zeta)^v - f_{N+1}(z+\zeta)^v| > c_1{}^v \quad \text{if } z \in L_1 \text{ and } |\zeta| = 1/v^2, \tag{5.3.3}$$

provided that v is sufficiently large. This will prove that no $z \in L_1$ can belong to a component of $\triangle_v$ which intersects K, and so the proof will be completed when (5.3.3) is verified.

To prove (5.3.3), we write

$$|f_j'(z+\zeta)^v - f_{N+1}(z+\zeta)^v| = |F_j(z+\zeta)^v - 1||f_{N+1}(z+\zeta)|^v.$$

Since $|f_{N+1}(z+\zeta)| \geq c_2(1 + O(v^{-2}))$, we have $|f_{N+1}(z+\zeta)|^v \geq c_2{}^v(1 + O(v^{-1})) > c_2{}^v/2$ if v is large. Now write

$$F_j(z+\zeta)^v - 1 = F_j(z)^v((F_j(z+\zeta)/F_j(z))^v - 1) + F_j(z)^v - 1,$$

and note that Taylor expansion gives

$$F_j(z+\zeta)/F_j(z) = 1 + l_j(\zeta) + O(v^{-4}),$$

where all the linear forms l_j do not vanish simultaneously since the functions F_j, $j = 1, \cdots, N$, have rank n on L_1. Hence $\max_{1 \leq j \leq N} |l_j(\zeta)| \geq c|\zeta|$

for some $c > 0$. We have $(F_j(z + \zeta)/F_j(z))^\nu = 1 + \nu l_j(\zeta) + O(\nu^{-2})$, so summing up the estimates above we obtain

$$\max_{1 \le j \le N} |f_j(z + \zeta)^\nu - f_{N+1}(z + \zeta)^\nu| > c_2{}^\nu 2^{-2}(c/\nu + O(\nu^{-2})) > c_1{}^\nu,$$
$$|\zeta| = 1/\nu^2,$$

if ν is large enough. The estimates are uniform in z when $z \in L_1$. This completes the proof of the lemma.

We can now prove the main result of this paragraph.

Theorem 5.3.9. *If Ω is a Stein manifold of dimension n, there exists an element $f \in A(\Omega)^{2n+1}$ which defines a one-to-one regular proper map of Ω into $\mathbf{C}^{2n+1}$.*

Proof. According to Theorem 5.3.6 there exists a regular one-to-one map g into $\mathbf{C}^{2n+1}$. If we can construct $f \in A(\Omega)^{2n+1}$ such that

$$\{z; z \in \Omega, |f(z)| \le k + |g(z)|\} \subset\subset \Omega \tag{5.3.4}$$

for every k, the theorem will follow. (Here we have written $|f(z)|$ for $\max_j |f_j(z)|$ and defined $|g(z)|$ similarly.) For application of Lemmas 5.3.4 and 5.3.5 to the regular one-to-one mapping (f,g) then shows that there exist matrices a_{jk} with constant, arbitrarily small coefficients, such that

$$f_j'(z) = f_j(z) + \sum_{k=1}^{2n+1} a_{jk} g_k, \qquad j = 1, \cdots, 2n + 1,$$

defines a one-to-one regular mapping f' into $\mathbf{C}^{2n+1}$. If $\Sigma_k |a_{jk}| \le 1$, we have

$$\{z; z \in \Omega, |f'(z)| \le k\} \subset \{z; z \in \Omega, |f(z)| \le k + |g(z)|\} \subset\subset \Omega,$$

so f' will also be proper.

To construct f we first note that, from (α) in Definition 5.1.3, it follows that there exists a sequence of compact subsets K_j of Ω such that K_j is in the interior of K_{j+1} for every j, $\hat{K}_j = K_j$ and $\cup_1^\infty K_j = \Omega$. By Lemmas 5.3.7 and 5.3.8 we can choose an analytic polyhedron P_j of order $2n$ such that $K_j \subset P_j \subset K_{j+1}$. Let

$$M_j = \sup_{P_j} |g|.$$

The condition (5.3.4) is then a consequence of the following:

$$|f| \ge k + M_{k+1} \quad \text{in } P_{k+1} \setminus P_k \text{ for every } k. \tag{5.3.5}$$

For (5.3.5) implies that $|f| \ge k + |g|$ in $P_{k+1} \setminus P_k$, hence $|f| \ge k + |g|$ in $\cup_k^\infty (P_{j+1} \setminus P_j) = \Omega \setminus P_k$.

We first construct $f_1, \cdots, f_{2n} \in A(\Omega)$ such that

(5.3.6) $$\max_{1 \le j \le 2n} |f_j(z)| > k + M_{k+1} \quad \text{on } \partial P_k \text{ for every } k.$$

To do so we note that by the definition of an analytic polyhedron of order $2n$ one can find $(h_1^k, \cdots, h_{2n}^k) \in A(\Omega)^{2n}$ so that $\max_j |h_j^k| < 1$ in $\overline{P_{k-1}}$ but $\max_j |h_j^k| = 1$ on ∂P_k. If we set $f_j^k = (a_k h_j^k)^{m_k}$ with a_k slightly larger than 1 and m_k a large integer, we can successively choose a_k and m_k so that for every k

$$\max_{1 \le j \le 2n} |f_j^k| \le 2^{-k} \text{ in } P_{k-1},$$

$$\max_{1 \le j \le 2n} |f_j^k| > M_{k+1} + k + 1 + \max_{1 \le j \le 2n} \left| \sum_{l=1}^{k-1} f_j^l \right| \text{ on } \partial P_k.$$

These conditions imply that the sums

$$f_j = \sum_{k=1}^{\infty} f_j^k, \qquad j = 1, \cdots, 2n,$$

converge to functions in $A(\Omega)$, and the construction immediately gives (5.3.6).

Now set

$$G_k = \{z; z \in P_{k+1} \setminus P_k, \max_{1 \le j \le 2n} |f_j(z)| \le k + M_{k+1}\},$$

$$H_k = \{z; z \in P_k, \max_{1 \le j \le 2n} |f_j(z)| \le k + M_{k+1}\}.$$

From (5.3.6) it follows that these disjoint sets are compact. The $A(\Omega)$-hull of $G_k \cup H_k$ is contained in K_{k+2} and can obviously be written $G_k \cup H_k \cup H_k'$ where $H_k' \subset \complement P_{k+1}$. (In fact, H_k' is empty but this has no importance for us.) Using Theorem 5.2.8 to approximate by functions in $A(\Omega)$ a function which is 0 on $H_k \cup H_k'$ and equal to some large constant on G_k, we obtain successively functions $h_k \in A(\Omega)$ such that

$$|h_k| < 2^{-k} \text{ in } H_k, \quad |h_k| \ge 1 + k + M_{k+1} + \Big|\sum_{j<k} h_j\Big| \text{ in } G_k, \quad k = 1, 2, \cdots.$$

Since $G_k \subset H_{k+1} \subset H_{k+2} \cdots$, the analytic function f_{2n+1} defined by

$$f_{2n+1} = \sum_{1}^{\infty} h_k$$

satisfies the inequality $|f_{2n+1}(z)| \ge k + M_{k+1}$, when $z \in G_k$. Hence (5.3.5) holds, which completes the proof.

5.4. Envelopes of holomorphy. In section 2.5 we proved that if Ω is a connected Reinhardt domain containing 0, or if Ω is a connected tube, there is a holomorphy domain $\tilde{\Omega}$ of the same type to which all functions which are analytic in Ω can be extended. We shall now give a general discussion of such results. It is not possible to restrict the study to open subsets of $\mathbf{C}^n$, so Stein manifolds naturally take the place of domains of holomorphy. Throughout this section we require all manifolds to be *connected* and countable at infinity without making this assumption explicitly in every statement.

We shall say that a manifold $\tilde{\Omega}$ is a *holomorphic extension* of another manifold Ω if

(i) Ω is an open subset of $\tilde{\Omega}$.
(ii) The analytic structure of Ω is induced by that in $\tilde{\Omega}$.
(iii) For every $u \in A(\Omega)$ one can find $\tilde{u} \in A(\tilde{\Omega})$ so that $u = \tilde{u}$ in Ω. (The function $\tilde{u}$ is then uniquely determined by u, for $\tilde{\Omega}$ is connected.)

We are interested in finding a Stein manifold $\tilde{\Omega}$ which is a holomorphic extension of a given manifold Ω. It is clear that Ω must then satisfy conditions (β) and (γ) of Definition 5.1.3.

Lemma 5.4.1. *If $\tilde{\Omega}$ is a holomorphic extension of Ω, one can for every compact subset $\tilde{K}$ of $\tilde{\Omega}$ find a compact subset K of Ω such that the $A(\tilde{\Omega})$-hull of K contains $\tilde{K}$.*

Proof. $A(\Omega)$ is a Fréchet space with the topology defined by all seminorms of the form

$$u \to \sup_K |u|, \qquad u \in A(\Omega),$$

where K is a compact subset of Ω. In fact, the topology is defined by countably many seminorms since Ω is countable at infinity, and the completeness follows from Corollary 2.2.4. Similarly, $A(\tilde{\Omega})$ is a Fréchet space. Now the fact that $\tilde{\Omega}$ is a holomorphic extension of Ω means that the restriction map

$$A(\tilde{\Omega}) \to A(\Omega)$$

is onto and, since it is continuous and one-to-one, the inverse is continuous by Banach's theorem. For every compact set $\tilde{K} \subset \tilde{\Omega}$, one can therefore find a compact set $K \subset \Omega$ and a constant C such that

$$\sup_{\tilde{K}} |\tilde{u}| \leq C \sup_K |\tilde{u}|, \qquad \tilde{u} \in A(\tilde{\Omega}).$$

Replacing $\tilde{u}$ by $\tilde{u}^k$, taking k^{th} roots and letting $k \to \infty$, we conclude that C can be chosen equal to 1. This proves the lemma.

Theorem 5.4.2. *If* Ω *is a Stein manifold, and* $\tilde{\Omega}$ *is a holomorphic extension of* Ω, *then* $\Omega = \tilde{\Omega}$.

Proof. If $\Omega \neq \tilde{\Omega}$, there must exist a boundary point $z \in \tilde{\Omega}$ of Ω, for $\tilde{\Omega} \setminus \Omega$ would otherwise be open, hence $\tilde{\Omega}$ would not be connected. Let $\tilde{K}$ be a compact neighborhood of z in $\tilde{\Omega}$. By Lemma 5.4.1, $\tilde{K} \cap \Omega$ is then in the $A(\Omega)$-hull of a compact subset of Ω, but this is impossible since Ω is a Stein manifold.

Stein manifolds are maximal not only in the sense that they have no holomorphic extensions, but also in the sense that, if one can find a holomorphic extension which is a Stein manifold, it contains all natural holomorphic extensions:

Theorem 5.4.3. *Let* Ω_1 *and* Ω_2 *be holomorphic extensions of* Ω, *and assume that* Ω_1 *is a Stein manifold and that functions in* $A(\Omega_2)$ *give local coordinates everywhere in* Ω_2 *and separate points in* Ω_2. *Then there is an analytic isomorphism* φ *of* Ω_2 *into* Ω_1 *which is the identity on* Ω; *if* Ω_2 *is a Stein manifold, it is an isomorphism onto.*

Hence there is apart from isomorphisms at most one holomorphic extension of Ω which is a Stein manifold. When such an extension exists, we call it the *envelope of holomorphy* of Ω.

Proof of Theorem 5.4.3. If $u \in A(\Omega)$, we denote by $E_j u$ the analytic continuation of u to Ω_j, $j = 1, 2$. When $z_2 \in \Omega_2$ and $z_1 \in \Omega_1$, we set $\varphi(z_2) = z_1$ if

$$(E_1 u)(z_1) = (E_2 u)(z_2) \quad \text{for every } u \in A(\Omega). \tag{5.4.1}$$

Since analytic functions separate points in both Ω_1 and Ω_2, this defines a one-to-one map of a subset of Ω_2 on a subset of Ω_1. The map φ is continuous and defined in a closed set. For let φ be defined at z_2^ν, $\nu = 1, 2, \cdots$ and let $z_2^\nu \to z_2$ in Ω_2. Then these points form a compact set $K_2 \subset \Omega_2$, so by Lemma 5.4.1 there is a compact set $K \subset \Omega$ with $A(\Omega_2)$-envelope containing K_2. Then $\varphi(z_2^\nu) \in \hat{K}_{\Omega_1}$ for every ν by (5.4.1) and, since Ω_1 is a Stein manifold, this is a compact set. Hence the sequence $\varphi(z_2^\nu)$ has a limit point z_1, and

$$(E_1 u)(z_1) = \lim_{\nu \to \infty} ((E_1 u)(\varphi(z_2^\nu)) = \lim_{\nu \to \infty} (E_2 u)(z_2^\nu) = (E_2 u)(z_2), \qquad u \in A(\Omega).$$

Thus $z_1 = \varphi(z_2)$, so the limit point is unique and the sequence $\varphi(z_2^\nu)$ convergent.

Next note that for any two points $z_1 \in \Omega_1$ and $z_2 \in \Omega_2$, one can find

$f^1, \cdots, f^n \in A(\Omega)$ so that $E_j f^1, \cdots, E_j f^n$ form a local system of coordinates at z_j, $j = 1, 2$. Indeed, given $j = 1$ or 2, we can choose $f_j^1, \cdots, f_j^n \in A(\Omega)$ having this property at z_j. But if $f^k = a_1 f_1^k + a_2 f_2^k$, then $E_j(f^k), k = 1, \cdots, n$, is a coordinate system at z_j except when (a_1, a_2) satisfies an algebraic equation (the vanishing of the Jacobian with respect to a coordinate system) and this equation is not identically fulfilled. Hence one can choose a_1 and a_2 so that one obtains a coordinate system at both points.

Now let M be the set of all z_2 in the domain of φ such that for every $u \in A(\Omega)$ all derivatives of $E_j u$ at z_j ($z_1 = \varphi(z_2)$) with respect to $E_j f^1, \cdots, E_j f^n$ are the same for $j = 1$ and 2 if $f^1, \cdots, f^n$ are chosen as above. It is obvious that M is closed. But M is also open. For the equations $(E_1 f^k)(\zeta_1) = (E_2 f^k)(\zeta_2), k = 1, \cdots, n$, give an analytic isomorphism between connected neighborhoods of z_2 and $z_1 = \varphi(z_2)$ if $z_2 \in M$, and considering power series expansions in the local coordinates $E_j f^k$, $k = 1, \cdots, n$, we find that these equations imply (5.4.1) at ζ_1 and ζ_2. Hence M is equal to Ω_2 and φ is an analytic isomorphism of Ω_2 into Ω_1. Since $\varphi(\Omega_2)$ is a Stein manifold if Ω_2 is a Stein manifold, it follows from Theorem 5.4.2 that $\varphi(\Omega_2)$ must then be equal to Ω_1.

We shall now give a sufficient condition for the existence of an envelope of holomorphy.

Definition 5.4.4. *A complex manifold Ω of dimension n is called a Riemann domain if analytic functions separate points in Ω and there is an analytic map*

$$\varphi : \Omega \to \mathbf{C}^n$$

which is everywhere regular, that is, locally an isomorphism.

One can of course imagine a Riemann domain as lying above $\mathbf{C}^n$. Another way of stating the hypothesis is that there exist n functions (n = dimension) which form a local system of coordinates at every point.

The main result of this section is the following theorem of Oka.

Theorem 5.4.5. *Every Riemann domain Ω has an envelope of holomorphy $\tilde{\Omega}$, and $\tilde{\Omega}$ is also a Riemann domain.*

We shall first extend Ω as far as possible by the classical method of forming power series expansions of the functions in $A(\Omega)$. We shall then use the results of section 5.2 to prove that the manifold so obtained is a Stein manifold.

Let $z \in \Omega$. The restriction of the map φ to a suitable neighborhood of z is by hypothesis an analytic isomorphism onto a neighborhood of

$\varphi(z)$. Let π be the inverse. We can then define the derivatives $\partial^\alpha u(z)$ when $u \in A(\Omega)$ by

$$\partial^\alpha u(z) = \partial^\alpha u(\pi(\zeta))_{\zeta=\varphi(z)}.$$

It is clear that $\partial^\alpha u \in A(\Omega)$. For a fixed z_0 we now form the power series expansion

$$\sum_\alpha (\zeta - \varphi(z_0))^\alpha \, \partial^\alpha u(z_0)/\alpha! \tag{5.4.2}$$

When $\zeta = \varphi(z)$ for some z sufficiently close to z_0, the sum converges and is equal to $u(z)$. Let r_{z_0} be the supremum of all r such that the power series for every $u \in A(\Omega)$ converges in $\{\varphi(z_0)\} + rD$, where

$$D = \{z\,; |z_j| < 1, j = 1, \cdots, n\},$$

and put $D_{z_0} = \{\varphi(z_0)\} + r_{z_0}D$. Then the series (5.4.2) defines an analytic function u_{z_0} in D_{z_0} for every function u which is analytic in Ω, and we have for every α

$$(\partial^\alpha u)_{z_0} = \partial^\alpha(u_{z_0}). \tag{5.4.3}$$

We shall now make a Riemann domain Ω_{z_0} out of the disjoint union $\Omega \cup D_{z_0}$. To do so we identify $z \in \Omega$ and $\zeta \in D_{z_0}$ if $u(z) = u_{z_0}(\zeta)$ for every $u \in A(\Omega)$. When $u = \varphi_j$, a coordinate of φ, it follows that $\varphi_j(z) = \zeta_j$, that is, $\varphi(z) = \zeta$. Hence the map φ extends to Ω_{z_0} if we define it as the identity on D_{z_0}. Further, if $z \in \Omega$ is identified with $\zeta \in D_{z_0}$, we obtain $u(w) = u_{z_0}(\varphi(w))$ for all w close to z, for

$$\begin{aligned} u(w) &= \sum (\varphi(w) - \varphi(z))^\alpha \, \partial^\alpha u(z)/\alpha! \\ &= \sum (\varphi(w) - \zeta)^\alpha \, \partial^\alpha u_{z_0}(\zeta)/\alpha! = u_{z_0}(\varphi(w)), \end{aligned}$$

where the second equality follows from (5.4.3) and the definition of the equivalence relation. It is now clear that Ω_{z_0} is a Hausdorff space with the strongest topology for which the natural maps of Ω and of D_{z_0} into Ω_{z_0} are continuous, and the extension of the function φ to Ω_{z_0} makes it a Riemann domain. Note that the natural maps $\Omega \to \Omega_{z_0}$ and $D_{z_0} \to \Omega_{z_0}$ are analytic isomorphisms. We identify Ω with its image, and Ω_{z_0} is then a holomorphic extension of Ω.

Now take a dense countable subset $z_0, z_1, \cdots$ of Ω and form successively $\Omega_{z_0}, (\Omega_{z_0})_{z_1}, \cdots$. These manifolds increase to a Riemann domain Ω' which is a holomorphic extension of Ω. To Ω' we apply the same method to obtain a Riemann domain $\Omega'' = (\Omega')'$ and so on. Let $\tilde{\Omega}$ be the limit of these Riemann domains. It is again a Riemann domain which extends

Ω holomorphically, but now we have $\tilde{\Omega}' = \tilde{\Omega}$. In fact, if for some $\tilde{z} \in \tilde{\Omega}$ the power series

$$\sum_{\alpha} (\zeta - \tilde{\varphi}(\tilde{z}))^{\alpha} \partial^{\alpha}\tilde{u}(\tilde{z})/\alpha!, \tag{5.4.4}$$

where $\tilde{u}$ is the analytic continuation of $u \in A(\Omega)$, converges in the polydisc $D_{\tilde{z}} = \{\tilde{\varphi}(\tilde{z})\} + rD$ for every $u \in A(\Omega)$, then there exists an open neighborhood $\tilde{D}_{\tilde{z}}$ of $\tilde{z}$ which is mapped homeomorphically by $\tilde{\varphi}$ on $D_{\tilde{z}}$. In fact, if $\tilde{z} \in \Omega^{(n)}$, we can by construction find $\tilde{D}_{\tilde{z}} \subset \Omega^{(n+1)}$.

For every $\tilde{z} \in \tilde{\Omega}$ we denote by $d(\tilde{z})$ the boundary distance of $\tilde{z}$ in $\tilde{\Omega}$, that is, the supremum of all r such that there is a neighborhood $\tilde{D}$ of $\tilde{z}$ which is mapped homeomorphically on $\{\tilde{\varphi}(\tilde{z})\} + rD$ by $\tilde{\varphi}$. By the arguments just given, r is the radius of the largest polydisc where (5.4.4) converges for every $u \in A(\Omega)$. The proofs of Theorems 2.5.4 and 2.6.5 can therefore be repeated in order to prove that $-\log d$ is plurisubharmonic in $\tilde{\Omega}$. (d is finite and continuous everywhere unless $\tilde{\Omega} = \mathbf{C}^n$.) Theorem 5.4.5 is thus a consequence of the following

Theorem 5.4.6. *Let Ω be a Riemann domain such that the boundary distance d is finite and $-\log d$ is plurisubharmonic in Ω. Then Ω is a Stein manifold.*

Note the close relation to Theorem 5.2.10. However, $-\log d$ is not necessarily smooth and $\{z; -\log d(z) < c\}$ need not be compact. Combination of Theorem 5.1.6 with the proof of Theorem 2.6.7 also shows that $-\log d$ must be plurisubharmonic if Ω is a Stein manifold.

Proof of Theorem 5.4.6. Let $\Omega_c = \{z; z \in \Omega, d(z) > c\}$. We claim that the sets Ω_c are Stein manifolds when $c > 0$ and have the Runge property with respect to one another. (Cf. Theorem 4.3.3 which obviously extends to Stein manifolds in view of Corollary 5.2.9.) To do so, we have to show that, if $0 < b < c$ and K is a compact subset of Ω_c, then $\hat{K}_{\Omega_b}$ is a compact subset of Ω_c, and therefore by Theorem 4.3.3 independent of b.

When $0 < \varepsilon < \varepsilon_0$, the set K lies in one component of Ω_ε which we denote by Ω_ε'. Fix a point $z_0 \in \Omega_\varepsilon'$ and let $\rho(z)$ be the distance from z_0 to z in Ω_ε' with respect to the element of arc length $|d\varphi(z)|$. The set $\{z; z \in \Omega_\varepsilon', \rho(z) < C\}$ is relatively compact for every C. In fact, if it is relatively compact for one value of C, it follows by application of the Borel–Lebesgue lemma that it is also relatively compact when C is replaced by $C + \varepsilon/2$. Choosing a function $\chi \in C_0^\infty(D)$ with $\chi \geq 0$ and $\int \chi \, d\lambda = 1$, we set

$$\rho_\delta(z) = \int \rho(\pi(\varphi(z) - \delta\zeta))\chi(\zeta)\, d\lambda(\zeta),$$

where π is the analytic inverse of φ which is defined in $\{\varphi(z)\} + d(z)D$, and maps $\varphi(z)$ on z. It is clear that $\rho_\delta \in C^\infty(\Omega'_{2\varepsilon})$ if $\delta < \varepsilon$. Writing $\rho \circ \pi = f$, which is a Lipschitz continuous function with Lipschitz constant 1, we have for w in a neighborhood of $\varphi(z)$

$$\rho_\delta(\pi w) = \int f(w - \delta\zeta)\chi(\zeta)\, d\lambda(\zeta).$$

If D is a first-order derivative, we obtain by differentiating under the integral sign and changing variables

$$D\rho_\delta(\pi w) = \int (Df)(\zeta)\chi((w - \zeta)/\delta)\, d\lambda(\zeta)/\delta^{2n}.$$

Since $|Df| \le 1$, it follows that $|D\rho_\delta(\pi w)| \le 1$ and that each derivative of $D\rho_\delta$ is bounded by a constant depending only on δ. Hence it is possible to choose a constant C_δ so that

$$\rho_\delta'(z) = \rho_\delta(z) + C_\delta \sum_1^n |\varphi_j(z)|^2$$

is strictly plurisubharmonic in $\Omega'_{2\varepsilon}$, and since $\rho - \rho_\delta \le \delta$, it is still true that $\{z\,;\, z \in \Omega'_{2\varepsilon}, \rho_\delta'(z) < C\}$ is relatively compact in Ω for every C.

Now let p_δ be the regularization of $-\log d$, defined similarly in Ω_δ. From Theorem 2.6.3 it follows that p_δ is plurisubharmonic. We have

$$-\log d \le p_\delta \le -\log(d - \delta).$$

Set

$$O = \{z\,;\, z \in \Omega_\delta, p_\delta(z) < -\log(2\varepsilon)\}.$$

Then O contains $\Omega_{3\varepsilon}$ since $0 < \delta < \varepsilon$, and O is contained in $\Omega_{2\varepsilon}$. Let $3\varepsilon < \varepsilon_0$ and $3\varepsilon < b$, and let O' be the component of O which contains K. Set in O'

$$q(z) = -(p_\delta(z) + \delta + \log c)/(p_\delta(z) + \log(2\varepsilon)).$$

If δ is small enough, we have $p_\delta(z) \le -\delta - \log c$ in K since $K \subset \Omega_c$. Thus $q \le 0$ in K; but $q(z) \to \infty$ if z approaches a boundary point of O'. Since q is a convex increasing function of p_δ, it is plurisubharmonic.

Now consider the function

$$\psi(z) = \rho_\delta'(z) + \lambda q(z),$$

where λ is a positive parameter. If C is the maximum of $\rho_\delta'(z)$ in K, then $\psi(z) \le C$ in K, and $\{z\,;\, z \in O', \psi(z) < \gamma\} \subset\subset O'$ for every γ. Indeed, $\rho_\delta' < \gamma + 1$ in this set, so it is relatively compact in Ω, and since $q < \gamma/\lambda$ the closure contains no boundary point of O'. Hence it follows from Theorem 5.2.10 that O' is a Stein manifold and that $\psi(z) \le C$ in $\hat{K}_{O'}$.

Since this is true for every $\lambda > 0$, it follows that $q \leq 0$ in $\hat{K}_{O'}$, hence $p_\delta(z) < -\log c$ in $\hat{K}_{O'}$. This implies that $-\log d(z) < -\log c$ in $\hat{K}_{O'}$, so that $\hat{K}_{O'}$ is a compact subset of Ω_c, and since O' contains the components of Ω_b which intersect K, it follows that $\hat{K}_{\Omega_b}$ is a compact subset of Ω_c. Hence all Ω_c are Stein manifolds and they have the relative Runge property.

Now let $K_1, K_2, \cdots$ be an increasing sequence of compact subsets of Ω such that every compact subset of Ω is contained in one of them, let $K_j \subset \Omega_{b_j}$, and assume that K_j is $A(\Omega_{b_j})$-convex, where b_j is a sequence of positive numbers $\to 0$. Any function which is analytic in a neighborhood of K_j can then be uniformly approximated on K_j by a function which is analytic in $\Omega_{b_{j+1}}$. This function can in turn be approximated arbitrarily closely on K_{j+1} by functions which are analytic in $\Omega_{b_{j+2}}$ and so on. Hence all functions which are analytic in a neighborhood of K_j can be uniformly approximated on K_j by functions which are analytic in Ω. It follows that the $A(\Omega)$-hull of each K_j is equal to K_j; hence Ω is a Stein manifold.

5.5. The Cousin problems on a Stein manifold. The results of section 5.2 make it easy to extend the Mittag–Leffler and Weierstrass theorems to Stein manifolds. We prefer to give the results in a form parallel to Theorem 1.4.5.

Theorem 5.5.1. *Let Ω be a Stein manifold and Ω_j open subsets of Ω such that $\Omega = \cup_1^\infty \Omega_j$. If $g_{jk} \in A(\Omega_j \cap \Omega_k)$, $j, k = 1, 2, \cdots$ and*

(5.5.1) $g_{jk} = -g_{kj}; g_{ij} + g_{jk} + g_{ki} = 0$ *in $\Omega_i \cap \Omega_j \cap \Omega_k$ for all i, j, k,*

then one can find functions $g_j \in A(\Omega_j)$ such that

(5.5.2) $g_{jk} = g_k - g_j$ *in $\Omega_j \cap \Omega_k$ for all j and k;*

in other words, the first Cousin problem has a solution.

Proof. We just have to repeat the proof of Theorem 1.4.5. Thus we choose a partition of unity and define functions h_k as there. Then $h_k \in C^\infty(\Omega_k)$ and

$$h_k - h_j = g_{jk} \quad \text{in } \Omega_j \cap \Omega_k.$$

This implies that $\bar\partial h_k = \bar\partial h_j$ in $\Omega_j \cap \Omega_k$, so there is a form $\psi \in C^\infty_{(0,1)}(\Omega)$ such that $\psi = \bar\partial h_k$ in Ω_k for every k. By Corollary 5.2.6, the equation $\bar\partial u = -\psi$ has a solution $u \in C^\infty(\Omega)$, and the functions $g_k = h_k + u$ then have all the required properties.

Next we consider the second Cousin problem, the analogue of the Weierstrass theorem. In doing so, we denote by $A^*(\Omega)$ the set of functions in $A(\Omega)$ which are *everywhere* different from 0, so that their reciprocals are also in $A(\Omega)$.

Theorem 5.5.2. *Let Ω be a Stein manifold and Ω_j open subsets of Ω such that $\Omega = \cup_1^\infty \Omega_j$. If $g_{jk} \in A^*(\Omega_j \cap \Omega_k)$, $j, k = 1, 2, \cdots$, and if*

$$g_{jk}g_{kj} = 1;\ g_{ij}g_{jk}g_{ki} = 1 \quad \text{in } \Omega_i \cap \Omega_j \cap \Omega_k \text{ for all } i, j, k, \tag{5.5.3}$$

then one can find $g_j \in A^*(\Omega_j)$ so that

$$g_{jk} = g_k g_j^{-1} \quad \text{in } \Omega_j \cap \Omega_k \text{ for all } j \text{ and } k, \tag{5.5.4}$$

provided that there exist nonvanishing functions $g_k \in C(\Omega_k)$ satisfying (5.5.4).

Proof. Let c_j be nonvanishing functions in $C(\Omega_j)$ such that

$$g_{jk} = c_k c_j^{-1}.$$

First assume that all the sets Ω_j are simply connected. Then we can write $c_j = e^{b_j}$, where $b_j \in C(\Omega_j)$. If we set $h_{jk} = b_k - b_j$, we obtain $g_{jk} = \exp h_{jk}$, so that h_{jk} is a unique continuous definition of the logarithm of g_{jk} and therefore analytic in $\Omega_j \cap \Omega_k$. We have

$$h_{ij} = -h_{ji};\ h_{ij} + h_{jk} + h_{ki} = 0 \quad \text{in } \Omega_i \cap \Omega_j \cap \Omega_k. \tag{5.5.5}$$

By Theorem 5.5.1 there exist functions $h_k \in A(\Omega_k)$ such that

$$h_{jk} = h_k - h_j \quad \text{in } \Omega_j \cap \Omega_k.$$

Writing $g_k = \exp h_k \in A^*(\Omega_k)$, we have then solved the second Cousin problem (5.5.4).

We now drop the assumption that the sets Ω_j are simply connected. Let $\{\Omega_\nu'\}_{\nu=1}^\infty$ be another covering of Ω with open simply connected sets Ω_ν' such that for every ν there is a positive integer i_ν for which $\Omega_\nu' \subset \Omega_{i_\nu}$. (That is, the covering $\{\Omega_\nu'\}$ is a refinement of the covering $\{\Omega_i\}$.) Set

$$g'_{\nu\mu} = g_{i_\nu i_\mu} \text{ in } \Omega_\nu' \cap \Omega_\mu'.$$

Then $g'_{\nu\mu}$ satisfies (5.5.3). If (5.5.4) holds for some continuous nonvanishing c_j, we obtain $g'_{\nu\mu} = c_\mu' c_\nu'^{-1}$ with $c_\nu' = c_{i_\nu}$. Hence the first part of the proof shows that one can find $g_\mu' \in A^*(\Omega_\mu')$ so that for all ν, μ

$$g'_{\nu\mu} = g_\mu' g_\nu'^{-1} \quad \text{in } \Omega_\nu' \cap \Omega_\mu'.$$

In particular, this implies that in $\Omega_i \cap \Omega_\nu' \cap \Omega_\mu' \subset \Omega_i \cap \Omega_{i_\nu} \cap \Omega_{i_\mu}$ we have

$$g_\mu' g_\nu'^{-1} g_{i_\mu i} g_{i i_\nu} = 1.$$

Here we have used (5.5.3). But this means that $g_\mu' g_{i_\mu i} = g_\nu' g_{i_\nu i}$ in $\Omega_\nu' \cap \Omega_\mu' \cap \Omega_i$, so there is a uniquely defined element $g_i \in A^*(\Omega_i)$ such that $g_i = g_\nu' g_{i_\nu i}$ in $\Omega_i \cap \Omega_\nu'$ for every ν. Now we obtain

$$g_k g_j^{-1} = g_\nu' g_{i_\nu k}(g_\nu' g_{i_\nu j})^{-1} = g_{i_\nu k} g_{j i_\nu} = g_{jk}$$

in $\Omega_\nu' \cap \Omega_j \cap \Omega_k$ for all ν, j, k. This proves (5.5.4).

Theorem 5.5.2 is a case of the Oka principle: on a Stein manifold it is "usually" possible to do analytically what one can do continuously. In section 7.4 we shall make the topological restriction in Theorem 5.6.2 explicit by interpreting it as the vanishing of a certain cohomology class. In particular, this condition is always fulfilled if $H^2(\Omega,\mathbf{Z}) = 0$ (and only then). An example of a Cousin problem which cannot be solved is therefore obtained by choosing a Stein manifold for which this group is not zero. Such an example was given after Theorem 4.2.7. For a direct discussion of an unsolvable Cousin problem, see Oka [3].

It is natural to ask if the topological difficulties which we encountered in Theorem 5.5.2 would disappear if the second Cousin problem is stated as follows (cf. Theorem 1.4.3′):

Given an open covering $\{\Omega_i\}$ of Ω and functions $f_i \in A(\Omega_i)$ such that $f_j/f_k \in A^*(\Omega_j \cap \Omega_k)$ for all j and k; find $f \in A(\Omega)$ so that $f/f_j \in A^*(\Omega_j)$ for every j.

If we set $g_{jk} = f_j/f_k$ in $\Omega_j \cap \Omega_k$ and $g_j = f/f_j$, this reduces to the Cousin problem as studied in Theorem 5.5.2. (See also the proof of Theorem 1.4.3′ by means of Theorem 1.4.5.) However, it is not obvious a priori that for arbitrary g_{jk} satisfying (5.5.3) one can find $f_j \in A(\Omega_j)$ so that $g_{jk} = f_j/f_k$. Thus the Cousin problem stated above might be more special than that considered in Theorem 5.5.2. The two statements are equivalent though, for we have the following theorem:

Theorem 5.5.3. *Let Ω_j, $j = 1, 2, \cdots$, be an open covering of a Stein manifold Ω, let $g_{jk} \in A^*(\Omega_j \cap \Omega_k)$ satisfy* (5.5.3). *Then one can find functions $f_j \in A(\Omega_j)$ not identically* 0 *so that*

(5.5.6) $$f_j = g_{jk} f_k \quad \text{in } \Omega_j \cap \Omega_k \text{ for all } j \text{ and } k.$$

Note that we do *not* assert that $f_j \in A^*(\Omega_j)$, so the theorem does *not* solve the second Cousin problem but merely asserts the equivalence of two different ways of posing it.

The proof of Theorem 5.5.3 will be given in the next section in a somewhat more general context.

5.6. Existence and approximation theorems for sections of an analytic vector bundle. Let Ω be a complex manifold. Then an analytic vector bundle B over Ω with N-dimensional fiber is an analytic manifold B together with

(i) an analytic map $p: B \to \Omega$ called the projection;
(ii) a vector space structure in each fiber $B_z = p^{-1}(z)$;

such that B is locally isomorphic to the product of an open subset of Ω and $\mathbf{C}^N$. This means that for each $z \in \Omega$ there exists an open neighborhood ω and an analytic mapping φ of $p^{-1}(\omega)$ onto $\omega \times \mathbf{C}^N$ such that φ^{-1} is analytic, and for every $z \in \omega$, φ maps B_z onto $\{z\} \times \mathbf{C}^N$ and the composed map

$$B_z \xrightarrow{\varphi} \{z\} \times \mathbf{C}^N \to \mathbf{C}^N$$

is linear, hence a linear isomorphism.

Let $\{\Omega_i\}_{i \in I}$ be an open covering of Ω such that for each i there is an analytic map φ_i of $p^{-1}(\Omega_i)$ onto $\Omega_i \times \mathbf{C}^N$ with the properties listed above. Then

$$g_{ij} = \varphi_i \varphi_j^{-1}$$

can be regarded as an analytic map of $\Omega_i \cap \Omega_j$ into the group $GL(N,\mathbf{C})$ of invertible $N \times N$ matrices with complex coefficients, and we have

$$\begin{aligned} g_{ij}g_{ji} &= \text{identity in } \Omega_i \cap \Omega_j, \\ g_{ij}g_{jk}g_{ki} &= \text{identity in } \Omega_i \cap \Omega_j \cap \Omega_k. \end{aligned} \tag{5.6.1}$$

A system of such $N \times N$ matrices g_{ij} with coefficients analytic in $\Omega_i \cap \Omega_j$ is called a system of transition matrices. We recall that the data in the second Cousin problem are precisely a system of transition functions.

The bundle B can be recovered from the system of transition matrices just defined. In fact, let B' be the set of all $(i, z, w) \in I \times \Omega \times \mathbf{C}^N$ such that $z \in \Omega_i$. We say that two elements (i, z, w) and (i', z', w') in B' are equivalent if $z = z'$ and $w' = g_{i'i}(z)w$. That this is an equivalence relation follows from (5.6.1). It is easy to verify that the space of equivalence classes, with the projection induced by the map $B' \ni (i, z, w) \to z$, is an analytic vector bundle for an arbitrary system of transition matrices.

If $\{g_{ij}\}$ is defined by means of a given analytic vector bundle, the construction gives back an isomorphic bundle.

If ω is an open subset of Ω, a C^∞ (or analytic) section of B over ω is by definition a C^∞ (or analytic) map

$$\omega \ni z \to u(z) \in B, \quad \text{such that } p(u(z)) = z.$$

If we have a covering $\{\Omega_i\}$ as above, this means that $\varphi_i \circ u = u_i$ is a C^∞ (or analytic) map of $\omega \cap \Omega_i$ into $\mathbf{C}^N$ such that

$$u_i = g_{ij} u_j \quad \text{in } \omega \cap \Omega_i \cap \Omega_j. \tag{5.6.2}$$

Conversely, any system of C^∞ (or analytic) maps u_i of $\omega \cap \Omega_i$ into $\mathbf{C}^N$ with these properties corresponds to precisely one C^∞ (or analytic) section of B over ω. To prove Theorem 5.5.3 therefore means to *construct a nontrivial analytic section of any analytic line bundle over a Stein manifold.*

We can also define the space of locally square integrable sections of B over ω to be a system of N-tuples u_i of locally square integrable functions in $\omega \cap \Omega_i$ satisfying (5.6.2). Similarly we can define distribution sections. We shall write $A(\omega,B)$, $C^\infty(\omega,B)$, $L^2(\omega,B,\text{loc})$, $W^s(\omega,B,\text{loc})$, $\cdots$ for the spaces of sections thus defined. Note that these definitions are independent of the choice of covering.

In order to carry over the L^2 methods we also have to define corresponding spaces of forms, for example $C^\infty_{(p,q)}(\omega,B)$. This we do by means of a covering $\{\Omega_i\}$ as above. To define an element $u \in C^\infty_{(p,q)}(\omega,B)$ thus means to give for every i an N-tuple u_i of forms in $C^\infty_{(p,q)}(\omega \cap \Omega_i)$ such that

$$u_i = g_{ij} u_j \quad \text{in } \omega \cap \Omega_i \cap \Omega_j.$$

(A different covering gives an isomorphic space.) Since g_{ij} is analytic, it follows that

$$\bar\partial u_i = g_{ij} \bar\partial u_j,$$

hence the N-tuples $\bar\partial u_j$ of forms of type $(p,q+1)$ define an element in $C^\infty_{(p,q+1)}(\omega,B)$. Similarly, the $\bar\partial$ operator is defined on $\mathscr{D}'_{(p,q)}(\omega,B)$.

Next we have to define hermitian norms on B-valued forms of type (p,q). To do so we first choose, as in section 5.2, a C^∞ hermitian metric in Ω so that (5.2.1) is valid. Then we choose a C^∞ hermitian metric in B, that is, a C^∞ function in B which restricted to each fiber B_z is a positive definite hermitian symmetric form. This can be done by means of a partition of unity. In a neighborhood U of any point in Ω we can by the Gram–Schmidt orthogonalization procedure construct C^∞ sections $b_1, \cdots, b_N$ of B such that $b_1(z), \cdots, b_N(z)$ form an orthonormal basis for B_z for every $z \in U$. Every B-valued form u of type (p,q) in U can then be written in one and only one way as a sum

$$u = \sum_1^N u^\nu b_\nu$$

where u^ν is a scalar form of type (p,q) in U. We set

$$|u|^2 = \sum_1^N |u^\nu|^2.$$

This definition is of course independent of the choice of the basis b_ν. The spaces $L^2_{(p,q)}(\Omega, B, \varphi)$ are now defined as in section 5.2, and so are the operators T and S from $L^2_{(p,q)}(\Omega, B, \varphi)$ to $L^2_{(p,q+1)}(\Omega, B, \varphi)$ and from $L^2_{(p,q+1)}(\Omega, B, \varphi)$ to $L^2_{(p,q+2)}(\Omega, B, \varphi)$, respectively. Since

$$|\bar\partial(\eta_\nu u) - \eta_\nu \bar\partial u|^2 = |\bar\partial\eta_\nu \wedge u|^2 \leq |\bar\partial\eta_\nu|^2\, |u|^2$$

by the definitions above, the proof of Lemma 5.2.1 is applicable with only formal changes to prove that $D_{(p,q+1)}(\Omega, B, \varphi)$ is dense in $D_S \cap D_{T^*}$ with respect to the graph norm. Furthermore,

$$\bar\partial u = \sum_1^N \bar\partial u^\nu b_\nu + \cdots$$

where dots indicate terms involving no differentiation of u. If $f \in C^\infty_{(p,q+1)}(U, B, \varphi)$ and we write $T^*f = \Sigma_1^N (T^*f)^\nu b_\nu$, it follows that $(T^*f)^\nu$ apart from terms involving no derivatives is given by (5.2.3) with f replaced by f^ν. Applying (5.2.10) to f^ν and adding for $\nu = 1, \cdots, N$, we conclude that (5.2.10) is valid for $f \in D_{(p,q+1)}(U,B)$. Repetition of the proof of Theorem 5.2.3 then again gives (5.2.12), and existence and approximation theorems follow as before:

Theorem 5.6.1. *Let Ω be a Stein manifold and B an analytic vector bundle over Ω. Then the equation $\bar\partial u = f$ has a solution $u \in W^{s+1}_{(p,q)}(\Omega,B)$ for every $f \in W^s_{(p,q+1)}(\Omega,B)$ such that $\bar\partial f = 0$. Every solution of the equation $\bar\partial u = f$ has this property when $q = 0$.*

Theorem 5.6.2. *Let Ω be a Stein manifold and φ a strictly plurisubharmonic function in Ω such that $K_c = \{z\,;\, z \in \Omega, \varphi(z) \leq c\}$ is compact for every real number c. Let B be an analytic vector bundle over Ω. Every analytic section of B over a neighborhood of K_0 can then be uniformly* *approximated (in the sense of the hermitian metric on B) by sections belonging to $A(\Omega,B)$.*

The proofs need not be repeated.

Since every point in a Stein manifold forms a holomorph-convex set, it follows from Theorems 5.1.6 and 5.6.2 that for every $z_0 \in \Omega$ and $b_0 \in B_{z_0}$ one can find sections $u \in A(\Omega,B)$ with $u(z_0)$ arbitrarily close to b_0. Hence one can find N analytic sections $u^1, \cdots, u^N$ of B over Ω such that $u^1(z_0), \cdots, u^N(z_0)$ are linearly independent, which implies

Corollary 5.6.3. *Let B be a vector bundle over a Stein manifold Ω. For every $z_0 \in \Omega$ and every $b_0 \in B_{z_0}$, one can find an analytic section u of B over Ω such that $u(z_0) = b_0$.*

In particular, this shows that the bundle defined by the transition functions of a second Cousin problem has a nontrivial section, that is, we have proved Theorem 5.5.3.

5.7. Almost complex manifolds. Let Ω be a C^∞ manifold of dimension $2n$. We shall say that Ω has an almost complex structure if we are given two mappings $P_{0,1}$ and $P_{1,0}$ of the space $C^\infty_{(1)}(\Omega)$ of complex valued first-order differential forms into itself such that

(i) $P_{0,1}$ and $P_{1,0}$ are linear over $C^\infty(\Omega)$.

(ii) $P_{0,1}$ and $P_{1,0}$ are complementary projections in the sense that

$$P_{0,1} + P_{1,0} = \text{identity}, \; P_{0,1}P_{1,0} = 0.$$

(iii) $P_{0,1}$ and $P_{1,0}$ are complex conjugate in the sense that

$$P_{0,1}\bar{f} = \overline{P_{1,0}f}, \qquad f \in C^\infty_{(1)}(\Omega).$$

Condition (i) implies that $P_{0,1}$ and $P_{1,0}$ induce linear mappings on the complexified cotangent space at every point. Thus these operators can be defined for differential forms over open subsets of Ω only. From conditions (ii) it follows that $P^2_{0,1} = P_{0,1}$, that $P^2_{1,0} = P_{1,0}$ and that $P_{1,0}P_{0,1} = 0$. In view of (iii), the operators $P_{0,1}$ and $P_{1,0}$ project the complexified cotangent space at each point on two complex conjugate n-dimensional subspaces.

Any complex structure in Ω defines in a natural way an almost complex structure where $P_{1,0}f$ and $P_{0,1}f$ are the parts of f spanned by differentials of analytic functions and their complex conjugates, respectively. The Cauchy–Riemann equations can then be written $P_{0,1}\,du = 0$, so the space of (local) analytic functions is determined by $P_{0,1}$. Hence there is at most one complex structure defining a given almost complex structure.

For any almost complex structure we shall write $\bar{\partial} u = P_{0,1}\, du$ if u is a differentiable function. The almost complex structure is then defined by an analytic structure if and only if for every point $x \in \Omega$ there exist C^∞ functions $u^1, \cdots, u^n$ in a neighborhood of x such that $\bar{\partial} u^j = 0$ for every j and $du^1, \cdots, du^n$ are linearly independent at x. The problem to decide when an almost complex structure is defined by an analytic structure is thus purely local, so we may assume that $\Omega \subset \mathbf{R}^{2n}$. Moreover, we may assume that there are n forms $\omega^1, \cdots, \omega^n$ of degree 1 with $P_{0,1}\omega^j = 0$ (we say that ω^j is of type (1,0) as in the analytic case) which are linearly independent at each point in Ω. Indeed, we can choose forms $f^1, \cdots, f^n$ such that $P_{1,0}f^1, \cdots, P_{1,0}f^n$ are linearly independent at any given point in Ω, and then the forms $\omega^j = P_{1,0}f^j$ have the desired property in some neighborhood. As in section 5.2, we can therefore write

$$du = \sum \partial u/\partial \omega^j \omega^j + \sum \partial u/\partial \bar{\omega}^j \bar{\omega}^j \tag{5.7.1}$$

as a definition of the linear first-order differential operators $\partial/\partial\omega^j$ and $\partial/\partial\bar{\omega}^j$. The equation $\bar{\partial} u = 0$ means that $\partial u/\partial\bar{\omega}^j = 0$ for all j. From (5.7.1) it follows that all first-order differential operators are linear combinations of the operators $\partial/\partial\omega^j$ and $\partial/\partial\bar{\omega}^j$. Using this fact we can extend Lemma 4.2.4. We write $D^\alpha = (\partial/\partial x_1)^{\alpha_1} \cdots (\partial/\partial x_n)^{\alpha_n}$.

Lemma 5.7.1. *Let Ω be an open set in $\mathbf{R}^{2n}$, let $u \in L^2(\Omega)$ have compact support in Ω, and assume that $\partial u/\partial\bar{\omega}^j \in L^2(\Omega)$, $j = 1, \cdots, n$. Then it follows that $u \in W^1(\Omega)$. If K is a compact subset of Ω, then*

$$\sum_{|\alpha| \le 1} \|D^\alpha u\|_{L^2} \le C(\|u\|_{L^2} + \sum_1^n \|\partial u/\partial\bar{\omega}^j\|_{L^2}) \quad \textit{if } \operatorname{supp} u \subset K. \tag{5.7.2}$$

Proof. We first prove (5.7.2) when $u \in C_0^\infty(K)$. Denote the adjoint of $\partial/\partial\bar{\omega}^j$ by δ_j. Then we have

$$\int |\partial u/\partial\omega^j|^2\, dx = \int |\partial u/\partial\bar{\omega}^j|^2\, dx + \int ((\delta_j \partial/\partial\bar{\omega}^j - \bar{\delta}_j \partial/\partial\omega^j)u)\bar{u}\, dx.$$

Since $\delta_j + \partial/\partial\omega^j$ is of order 0, the differential operator in the last integral is of order 1, so we obtain

$$\sum_{|\alpha|=1} \|D^\alpha u\|_{L^2}^2 \le C(\sum_1^n \|\partial u/\partial\bar{\omega}^j\|_{L^2}^2 + \|u\|_{L^2} \sum_{|\alpha|\le 1} \|D^\alpha u\|_{L^2}),$$

if we recall that all first-order operators are linear combinations of the operators $\partial/\partial\omega^j$ and $\partial/\partial\bar{\omega}^j$. The last term in this estimate can be replaced by

$$(n+1)C^2\|u\|_{L^2}^2 + \sum_{|\alpha|\le 1} \|D^\alpha u\|_{L^2}^2/2.$$

This gives (5.7.2) when $u \in C_0^\infty(K)$. Now if u only satisfies the hypotheses in the lemma, it follows from Lemma 5.2.2 with the notation used there that

$$\partial J_\varepsilon u/\partial\bar{\omega}^j - J_\varepsilon \partial u/\partial\bar{\omega}^j \to 0 \quad \text{in } L^2 \text{ when } \varepsilon \to 0.$$

Hence an application of (5.7.2) to $J_\varepsilon u - J_\delta u$ shows that $D^\alpha J_\varepsilon u$ is L^2 convergent when $\varepsilon \to 0$, which proves that $D^\alpha u \in L^2$ when $|\alpha| \le 1$.

Lemma 5.7.2. *Let Ω be an open set in $\mathbf{R}^{2n}$ and let $u \in L^2(\Omega,\text{loc})$. If $\partial u/\partial\bar{\omega}^j \in W^s(\Omega,\text{loc}), j = 1, \cdots, n$, for some positive integer s, it follows that $u \in W^{s+1}(\Omega,\text{loc})$. If K is a compact set in Ω and $\Omega' \subset \Omega$ is a neighborhood of K, then*

$$\sum_{|\alpha|\le s+1} \int_K |D^\alpha u|^2\,dx \le C\Big(\sum_{|\alpha|\le s} \sum_{j=1}^n \int_{\Omega'} |D^\alpha \partial u/\partial\bar{\omega}^j|^2\,dx + \int_{\Omega'} |u|^2\,dx\Big), \tag{5.7.3}$$

where C does not depend on u.

The proof follows from Lemma 5.7.1 by the arguments used in part (a) of the proof of Theorem 4.2.5, so we leave it as an exercise. For later reference we note that the Sobolev lemma shows (see the proof of Corollary 4.2.6) that u has continuous derivatives of order $s+1-2n$. Taking $s = 2n$ in (5.7.3), we obtain the estimate

$$\sum_{|\alpha|\le 1} \sup_K |D^\alpha u|^2 \le C\Big(\sum_{|\alpha|\le 2n} \sum_{j=1}^n \int_{\Omega'} |D^\alpha \partial u/\partial\bar{\omega}^j|^2\,dx + \int_{\Omega'} |u|^2\,dx\Big). \tag{5.7.4}$$

We shall now determine when an almost complex structure is defined by an analytic structure. Recall that forms f of degree 1 with $P_{0,1}f = 0$ (resp. $P_{1,0}f = 0$) are said to be of type (1,0) (resp. (0,1)). Since any form of degree 1 can be written in a unique way as a sum of a form of type (1,0) and a form of type (0,1), it follows that a form of degree p can be written in one and only one way as a sum of forms of type (a,b), $a \ge 0$, $b \ge 0$, $a + b = p$, where by a form of type (a,b) we mean a linear combination of exterior products of a forms of type (1,0) and b forms of type (0,1). In general, the differential df of a form f of type (p,q) can have components of type (a,b) for all a and b with $a + b = p + q + 1$. However, if the almost complex structure is defined by an analytic structure, we know that df is the sum of a form of type $(p+1,q)$ and another of type $(p,q+1)$. The condition in the following definition is therefore necessary for an almost complex structure to be defined by an analytic structure.

Definition 5.7.3. *An almost complex structure is called integrable if* df *has no component of type* (2,0) *when* f *is of type* (0,1).

If f is of type (1,0), it follows that $d\bar{f}$ has no component of type (2,0), that is, df has no component of type (0,2). Hence it follows that if f is of type (p,q), then df is a sum of a form ∂f of type $(p + 1,q)$ and another $\bar{\partial} f$ of type $(p,q + 1)$. In fact, it suffices to verify this for a form

$$f = g_1 \wedge \cdots \wedge g_p \wedge h_1 \wedge \cdots \wedge h_q$$

where $g_1, \cdots, g_p$ are of type (1,0) and $h_1, \cdots, h_q$ are of type (0,1), and then the statement follows from Definition 5.7.3 and the remark following it. Having now defined the operators ∂ and $\bar{\partial}$ in general, we conclude that (2.1.4) is valid, that is,

$$\partial^2 = \partial\bar{\partial} + \bar{\partial}\partial = \bar{\partial}^2 = 0.$$

For an integrable almost complex structure, we thus have the same formalism as for an analytic structure, and indeed we shall prove

Theorem 5.7.4. *Every integrable almost complex structure is defined by a unique analytic structure.*

Proof. The uniqueness is obvious, as was pointed out above. What we have to prove is therefore that for any $x \in \Omega$ there exist n C^∞ functions $u^1, \cdots, u^n$ in a neighborhood of x such that $\bar{\partial} u = 0$ and the differentials $du^1, \cdots, du^n$ are linearly independent at x. We may assume in the proof that $\Omega \subset \mathbf{R}^{2n}$, that $x = 0$, and that there are forms $\omega^1, \cdots, \omega^n$ as discussed above which are of type (1,0) and linearly independent at each point in Ω. Now the reader will recall that in section 5.2 we worked with such forms corresponding to an analytic structure. However, the fact that the forms ω^j were obtained from an analytic structure was used only in proving (5.2.7) which was a consequence of the identity $\partial\bar{\partial} w + \bar{\partial}\partial w = 0$. Since this is true in the integrable almost complex case too, the proofs of Theorems 5.2.3 and 5.2.4 remain valid then. To take advantage of this fact we have to find a function φ which is "plurisubharmonic with respect to the almost complex structure." To do so, we note that if $\psi(x) = |x|^2$, we have at 0

$$\sum \psi_{jk} t_j \bar{t}_k = \sum t_j\, \partial/\partial\varpi^j \overline{\sum t_k\, \partial/\partial\varpi^k} \sum_1^{2n} x_\nu^2 = 2 \sum_{\nu=1}^{2n} |\sum_k t_k\, \partial x_\nu/\partial\varpi^k|^2.$$

This is > 0 when $t \neq 0$, since the operators $\partial/\partial\varpi^k$ are linearly independent.

For a suitable $\delta > 0$, the ball $\Omega' = \{x; |x| < \delta\}$ is thus in Ω and $\Sigma\psi_{jk}t_j\bar{t}_k$ is uniformly positive definite there, so that $1/(\delta^2 - \psi)$ satisfies the hypotheses of Theorem 5.2.4 in Ω'. Hence we can choose φ equal to a convex increasing function of $1/(\delta^2 - \psi)$ so that, for every f of type (0,1) in Ω' with $\bar{\partial}f = 0$ and $\|f\|_\varphi < \infty$, there is a function u such that $\bar{\partial}u = f$ and

$$\|u\|_\varphi \leq \|f\|_\varphi. \tag{5.7.5}$$

To use this we shall choose v so that $f = \bar{\partial}v$ is nearly 0, then choose u so that $\bar{\partial}u = f$ and u is as small as f. Then $\bar{\partial}(u - v) = 0$, and $u - v$ should not be 0 if f is much smaller than v. Now we can only obtain good approximate solutions of the $\bar{\partial}$ equation very close to the origin, so we proceed as follows.

Let $u^1, \cdots, u^n$ be linear forms with $du^j = \omega^j$ at 0. Denote the mapping $x \to \varepsilon x$, $\varepsilon > 0$, by π_ε and consider the almost complex structure defined by the forms $\pi_\varepsilon{}^*\omega^1, \cdots, \pi_\varepsilon{}^*\omega^n$ in Ω'. All that we have proved above is true uniformly in ε when $0 < \varepsilon < 1$; and, since $du^j - \pi_\varepsilon{}^*\omega^j/\varepsilon$ is $O(\varepsilon)$ together with all its derivatives when $\varepsilon \to 0$, we conclude that

$$D^\alpha \bar{\partial}_\varepsilon u^j = O(\varepsilon) \quad \text{for all } \alpha,$$

if $\bar{\partial}_\varepsilon$ is the $\bar{\partial}$ operator with respect to the almost complex structure defined by the forms $\pi_\varepsilon{}^*\omega^k$. Hence it follows from (5.7.5) that we can find $v_\varepsilon{}^j$ so that $\bar{\partial}_\varepsilon v_\varepsilon{}^j = \bar{\partial}_\varepsilon u^j$ in Ω' and $\|v_\varepsilon{}^j\|_\varphi = O(\varepsilon)$. In view of (5.7.4) it follows that the derivatives of $v_\varepsilon{}^j$ at 0 are $O(\varepsilon)$. Hence the differentials of the functions $U^j = u^j - v_\varepsilon{}^j$ are linearly independent at 0 if ε is sufficiently small, and since $\bar{\partial}_\varepsilon U^j = 0$ it follows that the functions $U^j(x/\varepsilon)$ are solutions of the original $\bar{\partial}$ equation.

Notes. The class of manifolds which are now called Stein manifolds was first introduced by Stein [1]. In the seminars of Cartan [1], the theory of coherent analytic sheaves on Stein manifolds was developed (see Chapter VII). It contains the results of sections 5.2, 5.5, and 5.6 apart from the solution of the Levi problem (Theorem 5.2.10) which is due to Grauert [1]. For the origin of the methods we have used, see the notes to Chapter IV. They are not restricted to Stein manifolds; in fact, in Hörmander [1] they were used to prove more general results due to Andreotti and Grauert [1]. The embedding theorem in section 5.3 is due to Bishop [1] and Narasimhan [1]. We have mainly followed the proof of Bishop, who even proved that there is a proper map into $\mathbf{C}^{n+1}$. The results of section 5.4 are due to Oka [5]. The construction of the envelope of holomorphy can be done somewhat more elegantly by means of the notion of sheaf which we shall introduce in Chapter

VII. (See Cartan [1] and Malgrange [1].) However, the classical construction we give is basically the same apart from the fact that it may be too explicit. We refer to Bishop [2] for another proof that the envelope of holomorphy is a Stein manifold which does not rely on a solution of the Levi problem. The results on integrable almost complex structures proved in section 5.7 are due to Newlander and Nirenberg [1] but our proof is essentially that of Kohn [1]. For the applications to the study of perturbations of analytic structures we refer to a recent survey article by Nirenberg [1].

Chapter VI

LOCAL PROPERTIES OF ANALYTIC FUNCTIONS

Summary. After proving the Weierstrass preparation theorem we study divisibility properties in the ring of analytic functions at a point and results which are related to the fact that this ring is Noetherian. All the results are a preparation for the theory of coherent analytic sheaves in Chapter VII.

6.1. The Weierstrass preparation theorem. In this chapter we shall make a local study of analytic functions in $\mathbf{C}^n$, proceeding by induction with respect to n. The main point in the argument is the following extension of Corollary 1.2.10.

Theorem 6.1.1 (*The Weierstrass preparation theorem*). *Let f be analytic in a neighborhood ω of 0 in $\mathbf{C}^n$ and assume that $f(0,z_n)/z_n^p$ is analytic and $\neq 0$ at 0. Then one can find a polydisc $\Delta \subset \omega$ such that every g which is analytic and bounded in Δ can be written in the form*

$$g = qf + r, \tag{6.1.1}$$

where q and r are analytic in Δ, r is a polynomial in z_n of degree $< p$ (with coefficients depending on $z' = (z_1, \cdots, z_{n-1})$) and

$$\sup_\Delta |q| \leq C \sup_\Delta |g|, \tag{6.1.2}$$

where C is independent of g. The representation (6.1.1) *is unique. The coefficients of the power series expansions of q and r are finite linear combinations of those in the expansion of g.*

Proof. The hypothesis concerning f means that we can write

$$f = f_1 + z_n^p f_2,$$

where f_1 and f_2 are analytic in a neighborhood of 0, $f_2(0) \neq 0$, and f_1

is a polynomial in z_n of degree $< p$ which vanishes when $z' = 0$. Since $1/f_2$ is analytic in a neighborhood of 0, we can introduce $h = f_1 f_2^{-1}$ and $s = f_2 q$. Then we have $h(0,z_n) = 0$ identically and our task is to find s so that

$$g = (z_n^p + h)s + r. \tag{6.1.3}$$

Set

$$\Delta = \{z; |z_j| < R_j, j = 1, \cdots, n\}$$

with R_j chosen so small that f_1, f_2 and f_2^{-1} are analytic and bounded in Δ, and set $c = \sup_\Delta |h|$. Note that since $h(0,z_n) = 0$, we can for fixed R_n choose $R_1, \cdots, R_{n-1}$ so small that c is as small as we please.

Now we solve (6.1.3) by successive approximations. Thus we set $s_0 = 0$ and define s_k, r_k for $k \geq 1$ by the recursion formula

$$g = z_n^p s_k + h s_{k-1} + r_k, \tag{6.1.4}$$

where r_k shall be a polynomial in z_n of degree $< p$. Note that if φ is analytic in Δ and we write

$$\varphi = z_n^p \varphi_1 + \varphi_2,$$

where φ_2 is a polynomial in z_n of degree $< p$ and φ_1 is analytic, we must have

$$\varphi_2(z) = \sum_0^{p-1} \frac{\partial^j \varphi(z',0)}{\partial z_n^j} \frac{z_n^j}{j!}.$$

If $|\varphi| \leq M$ in Δ, it follows from Cauchy's inequalities that

$$|\varphi_2(z)| \leq pM, \qquad z \in \Delta,$$

for each term in the sum is bounded by M in absolute value. Hence $|\varphi_1(z) z_n^p| \leq (p + 1)M$, so the Schwarz lemma gives

$$|\varphi_1(z)| \leq (p + 1)M/R_n^p.$$

The recursion formula (6.1.4) thus determines uniquely the two sequences s_k and r_k, and from the fact that

$$z_n^p(s_{k+1} - s_k) + r_{k+1} - r_k = -h(s_k - s_{k-1}),$$

we obtain since $|h| \leq c$

$$\sup_\Delta |s_{k+1} - s_k| \leq c(p+1)R_n^{-p} \sup_\Delta |s_k - s_{k-1}|.$$

If Δ is chosen so that $c(p + 1) < R_n{}^p/2$, we conclude that

$$\sup |s_{k+1} - s_k| \leq 2^{-k} \sup |s_1|.$$

Hence

$$s = \lim s_k = \sum_1^\infty (s_k - s_{k-1})$$

exists with uniform convergence in Δ, and

$$|s| \leq 2 \sup_\Delta |s_1| \leq 2(p + 1)R_n{}^{-p} \sup_\Delta |g|.$$

Since the convergence of the sequence r_k now follows from (6.1.4), and since (6.1.4) converges to (6.1.3) when $k \to \infty$, we have obtained a solution of (6.1.3). This solution is unique, for if we have a solution of the homogeneous equation

$$(z_n{}^p + h)s + r = 0$$

such that s and r are bounded in Δ and r is a polynomial in z_n of degree $< p$, it follows that

$$\sup_\Delta |s| \leq c(p + 1)R_n{}^{-p} \sup_\Delta |s| \leq \tfrac{1}{2} \sup_\Delta |s|,$$

which implies that $s = 0$.

The last statement follows from the fact that $s_{k+1} - s_k$ and $r_{k+1} - r_k$ vanish to order k at 0.

Corollary 6.1.2. *If f satisfies the hypothesis of Theorem* 6.1.1, *then one can write f in one and only one way in the form $f = hW$, where h and W are analytic in a neighborhood of* 0, $h(0) \neq 0$, *and W is a Weierstrass polynomial, that is,*

$$W(z) = z_n{}^p + \sum_0^{p-1} a_j(z')z_n{}^j,$$

where a_j are analytic functions in a neighborhood of 0 *vanishing when* $z' = 0$.

Proof. This is the special case of Theorem 6.1.1 when $g(z) = z_n{}^p$. Note that, conversely, every f which can be represented as in the corollary must satisfy the hypotheses of Theorem 6.1.1.

For later reference we give two lemmas connecting divisibility of Weierstrass polynomials in the algebraic and in the analytic sense.

Lemma 6.1.3. *Let F, G, W be analytic in a neighborhood of* 0. *If* $F = GW$, *if F is a polynomial in* z_n *and W is a Weierstrass polynomial in* z_n, *then G is a polynomial in* z_n.

Proof. Since the leading coefficient of W, as a polynomial in z_n, is equal to 1, we can apply the algebraic division algorithm and obtain

$$F = G_1 W + H,$$

where H is a polynomial with respect to z_n of degree lower than that of W. But the uniqueness stated by the Weierstrass preparation theorem then implies that $G_1 = G$ and $H = 0$, and since G_1 is a polynomial, the lemma follows.

Note that the lemma may be false if W is an arbitrary polynomial with respect to z_n; for example, if W is a polynomial which does not vanish at 0, we may have $F = 1$ and $G = 1/W$.

Lemma 6.1.4. *Let F, G, W be analytic in a neighborhood of* 0, *W a Weierstrass polynomial, and F, G polynomials in* z_n. *If* $W = FG$, *it follows that F and G are also Weierstrass polynomials apart from a factor* $h(z')$ *with* $h(0) \neq 0$.

Proof. Let p be the degree of W; r and s the degrees of F and of G. Then $p = r + s$ and we have

$$z_n{}^p = W(0,z_n) = F(0,z_n)G(0,z_n).$$

Hence $F(0,z_n)/z_n{}^r$ and $G(0,z_n)/z_n{}^s$ are constants $\neq 0$, which means that the leading coefficients of F and of G do not vanish when $z' = 0$, but that all the others do.

Definition 6.1.5. *If f is analytic in a neighborhood of* 0, *we shall say that f is normalized in the* z_n*-direction if* $f(0,z_n)$ *does not vanish identically.*

With this terminology, the hypothesis of Theorem 6.1.1 means that f is normalized in the z_n-direction; the integer p is of course uniquely determined. It is obvious that by a linear change of variables we can always make any f which is not identically 0 normalized in the z_n-direction; we only have to choose the z_n axis so that it is not a zero of every homogeneous part of the Taylor expansion of f at 0.

6.2. Factorization in the ring A_0 of germs of analytic functions. Already in section 1.4 we stated the following definition in the case of one complex variable:

Definition 6.2.1. *If $z \in \mathbf{C}^n$ (or more generally z is in a complex manifold), we let $A_z(\mathbf{C}^n)$ or A_z for short denote the set of equivalence classes of functions f which are analytic in some neighborhood of z, under the equivalence relation $f \sim g$ if $f = g$ in some neighborhood of z. If f is analytic in a neighborhood of z, we write f_z or $\gamma_z(f)$ for the residue class of f in A_z, which is called the germ of f at z.*

When studying A_z it is of course no restriction to assume that $z = 0$. It is clear that A_0 is a ring with unit and without divisors of zero. Its elements can be identified with the set of all power series

$$\sum a_\alpha z^\alpha$$

which converge in some neighborhood of 0, that is, with the set of all arrays $\{a_\alpha\}$ such that $\Sigma\,|a_\alpha| r^{|\alpha|} < \infty$ for some $r > 0$. If $f \in A_0$, the value $f(0)$ at 0 is defined; it is equal to the constant term in the power series expansion. It is clear that f has an inverse in A_0 if and only if $f(0) \neq 0$. This means that the noninvertible elements in A_0 form an ideal (which thus contains all proper ideals).

The set of units in A_0, that is, the set of invertible elements, is the set of germs of functions which do not vanish at 0. If $f \in A_0$ and in every decomposition $f = gh$ with $f, g \in A_0$, one of the factors must be a unit, then f is called irreducible.

Theorem 6.2.2. *A_0 is a unique factorization domain, that is, every non-zero element in A_0 can be written as a product of irreducible factors in one and only one way—apart from units and the order of the factors.*

Proof. We recall from algebra that it suffices to prove the following two facts (see Zariski and Samuel [1], p. 21):

(a) If f is irreducible and f is a divisor of gh, then f divides either g or h.

(b) If in a sequence of elements $f_1, f_2, \cdots$ in A_0, each element is a divisor of the preceding one, then f_{k+1} differs from f_k only by a unit factor for large k.

In fact, (b) implies that f can be decomposed into a product of irreducible factors, and (a) implies that the decomposition is unique.

To prove the theorem we assume that it has already been proved for $n - 1$ variables and write $A_0' = A_0(\mathbf{C}^{n-1})$. In proving (a) we may also assume in view of Corollary 6.1.2 that f is the germ of a Weierstrass polynomial W (if necessary we have to change coordinates before applying Corollary 6.1.2). According to Theorem 6.1.1 we can also reduce g and h modulo f to polynomials g' and h' in $A_0'[z_n]$, and f divides $g'h'$ in $A_0'[z_n]$ by Lemma 6.1.3. Now f is irreducible as a polynomial in $A_0'[z_n]$ by Lemma 6.1.4. Since the inductive hypothesis implies that $A_0'[z_n]$ is a unique factorization domain (see, e.g., Zariski and Samuel [1], p. 32), this completes the proof of (a).

In proving (b) we may assume that f_1 is the germ of a Weierstrass polynomial. Each f_k must then be normalized in the z_n-direction and is therefore equivalent to a Weierstrass polynomial of degree n_k, which decreases with k. Hence the degree is constant for large k. But if a Weierstrass polynomial W divides another W' of the same degree, then $W = W'$, for if

$$W' = fW,$$

it follows that $f(0) = 1$ if we put $z' = 0$, and the uniqueness part of Corollary 6.1.2 then shows that $W = W'$. The proof is complete.

Theorem 6.2.3. *Let f and g be analytic in a neighborhood of 0 and assume that $\gamma_0(f)$ and $\gamma_0(g)$ are relatively prime. Then*

(i) *$\gamma_z(f)$ and $\gamma_z(g)$ are relatively prime in A_z for all z in a neighborhood of* 0.

(ii) *If $f(0) = g(0) = 0$, one can for every complex number a find z in any neighborhood of 0 so that $g(z) \neq 0$ and $f(z)/g(z) = a$.*

Thus it is not natural to assign any value, finite or infinite, to the quotient f/g at 0.

Proof. In the proof of (i) we may assume that both f and g are Weierstrass polynomials. If $A_0' = A_0(\mathbf{C}^{n-1})$, then f and g are relatively prime in $A_0'[z_n]$ (Lemma 6.1.4) and therefore in $K_0'[z_n]$ if K_0' is the quotient field of A_0' (Gauss' lemma; see Zariski and Samuel [1], p. 32). Hence we can find functions f_1, g_1, h which are analytic in a neighborhood of 0 so that f_1 and g_1 are polynomials in z_n, h is independent of z_n and not identically 0, and

$$h(z') = f_1(z)f(z) + g_1(z)g(z)$$

in a neighborhood of 0. Since the functions f and g are normalized with respect to z_n in the neighborhood of 0 where they are defined, a common factor of $\gamma_\zeta(f)$ and $\gamma_\zeta(g)$ in A_ζ must be normalized with respect to z_n and therefore can be assumed to be the germ of a Weierstrass polynomial at ζ. But since it must divide $\gamma_\zeta(h)$, it must be of degree 0 (see Lemma 6.1.3), which proves (i). In proving (ii) we may assume

$a = 0$, since we can otherwise replace f by $f - ag$, and we can then suppose that f and g are Weierstrass polynomials as above. Assume that the assertion is false, so that there is some neighborhood of 0 where $f(z) = 0$ implies $g(z) = 0$. Since the equation $f(z', z_n) = 0$ for fixed small z' has some small zero z_n, it follows that $h(z') = 0$ for all z' in a neighborhood of 0, which is a contradiction.

6.3. Finitely generated A_0-modules. We shall begin by strengthening part (b) of the proof of Theorem 6.2.2.

Definition 6.3.1. *A commutative ring A with unit is called Noetherian if every ideal $I \subset A$ is finitely generated, that is, if there exist elements $f_1, \cdots, f_j \in I$ so that every $f \in I$ can be written*

$$f = \sum_{1}^{j} a_i f_i$$

for some $a_i \in A$.

Lemma 6.3.2. *If A is Noetherian, then every submodule of A^p is finitely generated.*

By A^p we mean here the module of p-tuples of elements in A.

Proof. When $p = 1$ the assertion is identical to the definition of a Noetherian ring. Let M be any submodule of A^p and denote by π the projection of A^p on the first component, that is,

$$\pi(a_1, \cdots, a_p) = a_1.$$

Since πM is an ideal in A, it is finitely generated. Hence we can choose $m_1, \cdots, m_j \in M$ so that $\pi m_1, \cdots, \pi m_j$ generate πM. Every element $m \in M$ can therefore be written

$$m = a_1 m_1 + \cdots + a_j m_j + m_0,$$

where $a_1, \cdots, a_j \in A$ and $m_0 \in M$, $\pi m_0 = 0$. But

$$M_0 = \{m; m \in M, \pi m = 0\}$$

can be regarded as a submodule of A^{p-1}, so by induction with respect to p we may assume that M_0 is finitely generated. This completes the proof of the lemma.

Theorem 6.3.3. *A_0 is a Noetherian ring.*

Proof. The theorem is trivial when $n = 1$ since every ideal in A_0 is then generated by a power of z_1. Assume that the theorem has already been proved for the ring $A_0' = A_0(\mathbf{C}^{n-1})$. If I is an ideal in A_0 which contains some non-zero element, we can by a change of coordinates achieve that I contains the germ f of a function which is normalized with respect to the z_n-direction. Every $g \in I$ is then modulo f congruent to a polynomial in z_n of degree $< p$. Let M be the set of all $g \in I$ which are polynomials in z_n of degree $< p$. We can regard M as a submodule of $A_0'^p$ so M is finitely generated as an A_0'-module. The generators of M together with f then give a system of generators for the ideal I. The proof is complete.

We now introduce in A_0 the topology of simple convergence, defined by the countably many seminorms

$$f = \sum a_\alpha z^\alpha \to |a_\alpha|.$$

Thus $f_j \to f$ means that the coefficient of z^α in f_j converges to the coefficient of z^α in f for every α. Obviously A_0 is not a complete space; the completion is the space of all *formal power series* $\Sigma a_\alpha z^\alpha$. This is also a ring, which we denote by F_0. Having defined a topology in A_0 we have of course also a topology in $A_0{}^p$.

Theorem 6.3.4. *Let M be a submodule of $A_0{}^p$ and let $U_1, \cdots, U_j$ be generators of M. If $V_\nu \in M$ and $V_\nu \to 0$ when $\nu \to \infty$, one can find $f_\nu{}^i \in A_0$ so that*

$$V_\nu = \sum_1^j f_\nu{}^i U_i$$

and $f_\nu{}^i \to 0$ when $\nu \to \infty$, for every i.

Proof. We shall make an induction with respect to n and p, corresponding to the proofs of Lemma 6.3.2 and Theorem 6.3.3 combined.

(a) Assume that $p > 1$ and that the theorem has already been proved for submodules of $A_0{}^k$ when $k < p$. With π defined as in the proof of Lemma 6.3.2, we apply the inductive hypothesis to the sequence $\pi V_\nu \in \pi M$. It follows that we can find $f_\nu{}^i \in A_0$ such that $f_\nu{}^i \to 0$ when $\nu \to \infty$ and

$$V_\nu{}^0 = V_\nu - \sum_1^j f_\nu{}^i U_i \in M_0 \quad \text{for every } \nu.$$

Let $u_1, \cdots, u_r$ be generators for M_0. Then we can find $g_\nu{}^i \in A_0$ so that $g_\nu{}^i \to 0$ when $\nu \to \infty$ and

$$V_\nu^0 = \sum_1^r g_\nu{}^i u_i.$$

Since we can express each u_i as a linear combination with coefficients in A_0 of $U_1, \cdots, U_j$, the statement of the theorem follows.

(b) Now assume that $p = 1$ and that the theorem has already been proved for $n - 1$ variables and arbitrary p. We may assume that the ideal M contains a non-zero element f which is normalized in the z_n-direction. By the Weierstrass preparation theorem we can now write

$$V_\nu = q_\nu f + r_\nu,$$

where q_ν and $r_\nu \to 0$ and r_ν is in the set of polynomials in z_n of degree $< p$ with coefficients in $A_0' = A_0(\mathbf{C}^{n-1})$ that belong to M. Let $u_1, \cdots, u_s$ be a system of generators for this set considered as an A_0'-module. By the inductive hypothesis we then have

$$r_\nu = \sum_1^s g_\nu^i u_i$$

where $A_0' \ni g_\nu^i \to 0$ when $\nu \to \infty$. Since every u_i lies in the module generated by $U_1, \cdots, U_j$, the proof is now complete.

In exactly the same way we can prove

Theorem 6.3.5. *Every submodule of A_0^p is closed in A_0^p.*

Proof. (a) Assume that $p > 1$ and that the theorem has already been proved for submodules of A_0^k when $k < p$. Let $V_\nu \in M$ and $V_\nu \to V \in A_0^p$. By the inductive hypothesis, applied to the sequence πV_ν (notations of the proof of Lemma 6.3.2), it follows that $\pi V \in \pi M$. Let $U \in M$ and $\pi U = \pi V$. If $U_1, \cdots, U_j$ generate M, it follows from Theorem 6.3.4, since $\pi V_\nu - \pi U \to 0$, that we can choose $f_\nu^i \in A_0$ so that $f_\nu^i \to 0$ and

$$V_\nu^0 = V_\nu - U - \sum_1^j f_\nu^i U_i \in M_0.$$

Since $V_\nu^0 \to V - U$, it follows from the inductive hypothesis that $V - U \in M_0$, hence that $V \in M$.

(b) Now suppose that $p = 1$ and that the theorem has already been proved for $n - 1$ variables. We may assume that the ideal M contains a non-zero element f which is normalized in the z_n-direction. By the Weierstrass preparation theorem we can write

$$V_\nu = q_\nu f + r_\nu, \qquad V = qf + r,$$

and $q_\nu \to q$, $r_\nu \to r$ when $\nu \to \infty$. Thus it follows from the inductive hypothesis that r belongs to the submodule of $A_0'^p$ formed by polynomials in z_n of degree $< p$ with coefficients in A_0' which belong to M. Hence $V \in M$, which completes the proof.

Corollary 6.3.6. *Let $m_0, \cdots, m_j \in A_0{}^p$ and assume that*

$$m_0 = \sum_1^j f^i m_i \tag{6.3.1}$$

for some formal power series f^i. Then the same equality holds for some $f^i \in A_0$.

Proof. Let $f_v{}^i$ be the germ of the sum of the terms of degree $\le v$ in the formal power series f^i. Then $m_0 - \Sigma_1^j f_v{}^i m_i \to 0$ when $v \to \infty$; so it follows from Theorem 6.3.5 that m_0 belongs to the A_0-module generated by $m_1, \cdots, m_j$.

The interest of this result is that, at least in principle, it is elementary to decide if (6.3.1) holds for some $f^i \in F_0$. Indeed, this means to examine if an infinite system of linear equations has a solution, and we have

Lemma 6.3.7. *Let $L_1, L_2, \cdots$ be linear forms with complex coefficients in the complex variables $\xi_1, \xi_2, \cdots$, each form depending only on a finite number of the variables ξ_j. Then a system of equations*

$$L_j(\xi) = b_j, \qquad j = 1, 2, \cdots$$

with complex b_j has a solution ξ if every finite set of these equations is compatible.

Proof. The hypothesis means that if a finite sum $\Sigma\, c_j L_j$ is identically 0, then $\Sigma\, c_j b_j = 0$. We now consider two cases:

(a) If some finite sum $\Sigma\, c_j L_j$ is equal to ξ_1, then we can form a new system which is equivalent to the original one, where the first equation is $\xi_1 = b_1$ and the others do not involve ξ_1.

(b) If no finite linear combination $\Sigma\, c_j L_j$ is equal to ξ_1, then $\Sigma\, c_j L_j(\xi) = 0$ for all ξ if this sum vanishes when $\xi_1 = 0$. Hence we may take $\xi_1 = 0$ and obtain a new compatible system for $\xi_2, \cdots$.

In both cases we can thus choose some value for ξ_1 so that the system is reduced to a system of equations in $\xi_2, \cdots$ which satisfies the same hypotheses. Hence we can successively choose ξ_1, ξ_2, so that the infinite system of equations is satisfied, for any one of the given equations will be satisfied when we have chosen all the variables occurring in it.

6.4. The Oka theorem. As yet we have only considered A_z modules for a fixed z. However, in the next chapter we need the following result which goes beyond the Noetherian property of A_z by letting z be variable.

Theorem 6.4.1. *Let Ω be an open set in $\mathbf{C}^n$ and let $F_1, \cdots, F_q \in A(\Omega)^p$,*

$$R_z(F_1, \cdots, F_q) = \{G = (g^1, \cdots, g^q) \in A_z^{\,q};\ \sum_1^q g^j \gamma_z(F_j) = 0\}, z \in \Omega.$$

This is a submodule of $A_z^{\,q}$, called the module of relations between $F_1, \cdots, F_q$ at z. For any point in Ω one can then find an open neighborhood $\omega \subset \Omega$ and finitely many elements $G_1, \cdots, G_r \in A(\omega)^q$ so that R_z for every $z \in \omega$ is equal to the A_z-module which is generated by $\gamma_z(G_1), \cdots, \gamma_z(G_r)$.

Since A_z is Noetherian, we know of course already that R_z is finitely generated for every z, but the important point in the theorem is that one can use "the same" generators for all z in a neighborhood of any given point.

Proof of Theorem 6.4.1. We shall proceed by induction in essentially the same way as in section 6.3. It is of course sufficient to assume that $0 \in \Omega$ and construct a neighborhood ω of 0 with the required properties.

(a) Assume that $p > 1$ and that the theorem has already been proved for smaller values of p. Write $F_j = (F_j^{\,1}, \cdots, F_j^{\,p})$. It is obvious that

$$R_z(F_1, \cdots, F_q) \subset R_z(F_1^{\,1}, \cdots, F_q^{\,1}).$$

By the inductive hypothesis we can find a neighborhood $\Omega' \subset \Omega$ of 0 and finitely many elements $H_1, \cdots, H_r \in A(\Omega')^q$ so that $\gamma_z(H_1), \cdots, \gamma_z(H_r)$ generate the A_z module $R_z(F_1^{\,1}, \cdots, F_q^{\,1})$ for every $z \in \Omega'$. If $z \in \Omega'$, we therefore have

$$R_z(F_1, \cdots, F_q) \subset \left\{\sum_1^r c^j \gamma_z(H_j);\ c^j \in A_z\right\}.$$

Now the condition $\Sigma_1^r\, c^j \gamma_z(H_j) \in R_z(F_1, \cdots, F_q)$ means explicitly that

$$\sum_{j=1}^{r} \sum_{k=1}^{q} c^j \gamma_z(H_j^{\,k} F_k^{\,i}) = 0, \qquad i = 1, \cdots, p. \tag{6.4.1}$$

However, the choice of the elements H_j guarantees that the equation (6.4.1) is satisfied when $i = 1$. In reality we therefore only have $p - 1$ equations (6.4.1). By the inductive hypothesis it follows that there is a neighborhood $\omega \subset \Omega'$ of 0 and elements $C_1, \cdots, C_s \in A(\omega)^r$ so that the module of all $(c^1, \cdots, c^r) \in A_z^{\,r}$ satisfying (6.4.1) is generated by

$\gamma_z(C_1), \cdots, \gamma_z(C_s)$ if $z \in \omega$. The s elements

$$G_i = \sum_1^r C_i^j H_j, \qquad i = 1, \cdots, s,$$

therefore have all required properties.

(b) Now assume that $p = 1$ and that the theorem has already been proved for every p in the $(n-1)$-dimensional case. (The theorem is trivial when $n = 0$.) By a linear transformation we can achieve that the analytic functions $F_1, \cdots, F_q$ are normalized with respect to the z_n-direction, and then there is no restriction in assuming that they are all Weierstrass polynomials, with coefficients $\in A(\Omega')$, where Ω' is a neighborhood of 0 in $\mathbf{C}^{n-1}$. If $z' \in \Omega'$, we write $A'_{z'}$ for $A_{z'}(\mathbf{C}^{n-1})$. To proceed we need a lemma.

Lemma 6.4.2 *If $\zeta = (\zeta',\zeta_n)$ and $\zeta' \in \Omega'$, the A_ζ-module $R_\zeta(F_1, \cdots, F_q)$ is generated by those of its elements whose components are germs of functions in $A'_{\zeta'}[z_n]$ with a degree with respect to z_n which does not exceed μ, the maximum of the degrees of $F_1, \cdots, F_q$.*

Proof. Let F_q be of degree μ with respect to z_n,

$$F_q = z_n^\mu + \cdots$$

where the coefficients are in $A(\Omega')$. If $\zeta = (\zeta',\zeta_n)$ and $\zeta' \in \Omega'$, we can by the preparation theorem write

$$\gamma_\zeta(F_q) = F'F''$$

where F' and $F'' \in A_\zeta$, F' is the germ of a Weierstrass polynomial in $z - \zeta$ and $F''(\zeta) \neq 0$. From Lemma 6.1.3 it follows that F'' is a polynomial in z_n with leading coefficient equal to 1. Let μ' and μ'' denote the degrees of F' and of F'' with respect to z_n. If $(c^1, \cdots, c^q) \in R_\zeta(F_1, \cdots, F_q)$, we can by the preparation theorem write

$$c^i = \gamma_\zeta(F_q)t^i + r^i, \qquad i = 1, \cdots, q-1,$$

where t^i, $r^i \in A_\zeta$ and r^i is the germ of a polynomial in z_n of degree $< \mu'$ with coefficients in $A'_{\zeta'}$. With $r^q = c^q + \Sigma_{i=1}^{q-1} \gamma_\zeta(F_i)t^i$, we now write

$$(6.4.2) \qquad \begin{aligned}(c^1, \cdots, c^q) &= \gamma_\zeta(F_q,0, \cdots, 0, -F_1)t^1 \\ &+ \cdots + \gamma_\zeta(0, \cdots, 0, F_q, -F_{q-1})t^{q-1} + (r^1, \cdots, r^q).\end{aligned}$$

We must have $(r^1, \cdots, r^q) \in R_\zeta(F_1, \cdots, F_q)$, since all other terms in (6.4.2) are in that module. Thus

$$\sum_{1}^{q-1} r^i \gamma_\zeta(F_i) + (r^q F'')F' = 0.$$

Since the sum is a polynomial in z_n of degree $< \mu + \mu'$, it follows from Lemma 6.1.3 that $r^q F''$ is a polynomial in z_n of degree $< \mu$. Now we have

$$(r^1, \cdots, r^q) = 1/F''(F''r^1, \cdots, F''r^q)$$

where $F''r^j$ is of degree $< \mu$ for every j. In combination with (6.4.2) this proves the lemma.

End of proof of Theorem 6.4.1. If $c = (c^1, \cdots, c^q)$ is one of the elements in $R_\zeta(F_1, \cdots, F_q)$ described in Lemma 6.4.2, we can write

$$c^j = \sum_{k=0}^{\mu} c^{jk} \gamma_\zeta(z_n{}^k), \qquad c^{jk} \in A'_{\zeta'}.$$

The condition for $(c^1, \cdots, c^q)$ to belong to $R_\zeta(F_1, \cdots, F_q)$ therefore consists of $2\mu + 1$ linear homogeneous conditions in the quantities c^{jk}, with coefficients in $A(\Omega')$. By the inductive hypothesis one can therefore find a neighborhood $\omega' \subset \Omega'$ of 0 in $\mathbf{C}^{n-1}$ and elements $C_1, \cdots, C_r \in A(\omega')[z_n]^q$ of degree $\leq \mu$ with respect to z_n which for every $\zeta' \in \omega'$ generate the $A'_{\zeta'}$ module of solutions of these equations. From the lemma it now follows that the germs of $C_1, \cdots, C_r$ for every $\zeta = (\zeta', \zeta_n)$ with $\zeta' \in \omega'$ are generators for the A_ζ-module $R_\zeta(F_1, \cdots, F_q)$. This completes the proof.

Notes. The Oka theorem proved in section 6.4 first appeared in Oka [4]. The theory of coherent analytic sheaves could not be presented without it. However, in view of the emphasis of this book on analysis we have proved a minimum of results in this chapter. For example, no discussion of analytic sets is given here, so we refer to Hervé [1] or a forthcoming book by Gunning and Rossi for a much more extensive study of the local theory.

Chapter VII

COHERENT ANALYTIC SHEAVES ON STEIN MANIFOLDS

Summary. In this chapter the study of the Cousin problems is extended to coherent analytic sheaves on Stein manifolds. Definitions of sheaves, analytic sheaves, and coherent analytic sheaves are given in section 7.1. Loosely stated, they mean that sections of a coherent analytic sheaf can be described locally for some integer p as p-tuples of analytic functions modulo a finitely generated submodule of such p-tuples. Locally it is therefore easy to extend a number of results about analytic functions to theorems concerning sections of coherent analytic sheaves. The main new problem is thus to find global results. The key result is the existence of global sections of coherent analytic sheaves on Stein manifolds proved in section 7.2 (Theorem A of Cartan). Next we discuss the first Cousin problem with analytic functions replaced by sections of a coherent analytic sheaf. This extension requires a generalization of the Cousin problems involving the notion of cohomology group with values in a sheaf, which we introduce in section 7.3. (The reader is referred to Serre [1] for a more detailed discussion of the basic properties of sheaves and cohomology groups.) That the first Cousin problem can be solved means in this terminology precisely that the first cohomology group vanishes. In section 7.4 we first prove the vanishing of all the cohomology groups of positive order of the sheaf of germs of analytic functions. The fact that we consider cohomology groups of all orders makes it easy to extend this result to coherent analytic sheaves, using the existence of global sections. This gives Theorem B of Cartan. In section 7.5 we give the classical theorem of de Rham concerning the connection between the cohomology groups with complex coefficients and the existence of closed differential forms. Finally, in section 7.6 we return to the discussion of coherent analytic sheaves. Restricting ourselves to sheaves over $\mathbf{C}^n$ with polynomial generators, we establish a quantitative version of Theorem B. More specifically, we study systems of equations of the form $Pu = f$, where P is a rectangular matrix with polynomial entries, and u, f are s-tuples and t-tuples of analytic functions. Theorem B implies that there exists an analytic solution u for given analytic f if a solution exists locally. The results of section 7.6 give in addition a solution u which, roughly speaking, has the same growth as f. The proof is parallel to the proof of Theorem B but requires careful estimates at every step, so we have to rely on the results of section 4.4 and a re-examination of the local theory of Chapter VI. As an application we sketch briefly a proof of existence theorems for overdetermined systems of partial differential equations with constant coefficients.

7.1. Definition of sheaves. Let Ω be a complex manifold. As explained in section 6.2, we then have for every $z \in \Omega$ a ring A_z of germs at z of functions analytic in a neighborhood of z. Let

$$\mathscr{A} = \bigcup_{z \in \Omega} A_z$$

where the rings A_z are considered as disjoint sets. For every open subset ω of Ω and every $f \in A(\omega)$, we obtain a map

$$\omega \ni z \to \gamma_z(f) \in \mathscr{A} \tag{7.1.1}$$

which of course determines f uniquely, since $f(z) = (\gamma_z(f))(z)$. If π is the map $\mathscr{A} \to \Omega$ which maps A_z on z, the composition of the map (7.1.1) with π is the identity on ω. We wish to introduce a topology in $\mathscr{A}$ so that all maps of the form (7.1.1) with $f \in A(\omega)$ are continuous; and conversely every continuous map φ of an open set $\omega \subset \Omega$ into $\mathscr{A}$ with $\pi\varphi =$ identity on ω is defined according to (7.1.1) by a function $f \in A(\omega)$.

Assume that we have a topology in $\mathscr{A}$ so that the map (7.1.1) is continuous for every analytic f. Let U be an open set in $\mathscr{A}$ and let $F \in U$. The definition of $\mathscr{A}$ means that there exists a point $z \in \Omega$ and a function f which is analytic in a neighborhood ω of z such that $\gamma_z(f) = F$. If the map (7.1.1) is continuous, we must therefore be able to find a neighborhood $\omega' \subset \omega$ of z so that

$$\{f,\omega'\} = \{\gamma_\zeta(f); \zeta \in \omega'\} \subset U.$$

Hence every open set in $\mathscr{A}$ is a union of sets of the form $\{f,\omega\}$ with $f \in A(\omega)$ and ω open; conversely, all maps (7.1.1) are continuous if the topology in $\mathscr{A}$ has this property.

If $f' \in A(\omega')$ and $f'' \in A(\omega'')$, where ω' and ω'' are open, it follows that

$$\{f',\omega'\} \cap \{f'',\omega''\} = \{f',\omega\} = \{f'',\omega\},$$

where

$$\omega = \{z; z \in \omega' \cap \omega'', \gamma_z(f') = \gamma_z(f'')\}$$

is an open set. Hence the sets $\{f,\omega\}$ where $f \in A(\omega)$ and ω is an open set in Ω can be used as a basis for open sets to define a topology in $\mathscr{A}$; the open sets in $\mathscr{A}$ are then by definition all unions of sets of the form $\{f,\omega\}$.

This is the strongest topology for which all the maps (7.1.1) are continuous. It fulfills the Hausdorff separation axiom. In fact, if $f_z \neq g_w$, we have either $z \neq w$, which obviously implies that f_z and g_w have disjoint neighborhoods, or else $z = w$. If ω is a connected neighborhood of z where f and g are analytic, it follows that the neighborhoods $\{f,\omega\}$ and $\{g,\omega\}$ of $\gamma_z(f)$ and $\gamma_z(g)$ are disjoint. For f and g would otherwise define the same germ at some point in ω, which implies that $f = g$ in ω by the principle of analytic continuation; in particular, $f_z = g_z$, which is a contradiction. Finally note that, if ω is open and $f \in A(\omega)$, the restriction of π to $\{f,\omega\}$ is a homeomorphism onto ω, which is the inverse of (7.1.1).

Now let ω be an open set in Ω and φ a map $\omega \to \mathscr{A}$ with $\pi\varphi =$ identity which is continuous in the topology just defined. Let $f(z)$ be the value of the germ $\varphi(z) \in A_z$ at z. Then $f \in A(\omega)$ and $\varphi(z) = \gamma_z(f)$ for every $z \in \omega$. In fact, the definition of the topology in $\mathscr{A}$ means that to every point in ω there is a neighborhood ω' and an analytic function g so that $\varphi(z) \in \{g,\omega'\}$, that is, $\varphi(z) = \gamma_z(g)$, for every $z \in \omega'$. But this implies that $g = f$ in ω'.

With the topology we have introduced, the analytic functions can thus be identified with sections of $\mathscr{A}$ if we copy the definition of sections of bundles used in section 5.6. The space $\mathscr{A}$ is an example of a sheaf; we take its characteristic features as a definition:

Definition 7.1.1. *Let X and $\mathscr{F}$ be two topological spaces (not necessarily Hausdorff spaces) and let π be a mapping $\mathscr{F} \to X$ such that*

(i) *π maps $\mathscr{F}$ onto X.*
(ii) *π is a local homeomorphism, that is, every point in $\mathscr{F}$ has an open neighborhood which is mapped homeomorphically by π on an open set in X.*

Then $\mathscr{F}$ is called a sheaf on X and π is called the projection on X. If U is a subset of X, a section of $\mathscr{F}$ over U is a continuous map $\varphi: U \to \mathscr{F}$ such that $\pi\varphi =$ identity on U. The set of all sections of $\mathscr{F}$ over U is denoted by $\Gamma(U,\mathscr{F})$. If $x \in X$, then $\mathscr{F}_x = \pi^{-1}\{x\}$ is called the stalk of $\mathscr{F}$ at x.

Note that it follows from (ii) that $\varphi(U)$ is open in $\mathscr{F}$ if U is open in X and $\varphi \in \Gamma(U,\mathscr{F})$, and that the open sets in $\mathscr{F}$ are unions of sets of this type. Further note that if φ and $\psi \in \Gamma(U,\mathscr{F})$ and if $\varphi(z) = \psi(z)$ for some $z \in U$, then $\varphi(\zeta) = \psi(\zeta)$ for all $\zeta \in U$ in a neighborhood of z. Every element in $\mathscr{F}$ can therefore be regarded as a germ of sections of $\mathscr{F}$. If φ is a section we often write φ_z instead of $\varphi(z)$.

Examples. (1) The sheaf $\mathscr{A}$ of germs of analytic functions on a complex manifold Ω is a sheaf on Ω.

(2) On a C^∞ manifold we can in the same way define the sheaf of germs of C^∞ functions, usually denoted by $\mathscr{E}$, so that the sections of $\mathscr{E}$ are the C^∞ functions. In this case the Hausdorff axiom is not fulfilled, for if $f \in C^\infty$ in a neighborhood of x and if x is a boundary point of the complement of the support of f, then every neighborhood of f_x intersects every neighborhood of 0_x.

(3) If F is a topological space with the discrete topology, then $\mathscr{F} = X \times F$ is a sheaf (called a constant sheaf).

The sheaves in which we are interested carry additional algebraic structure. We shall say that $\mathscr{F}$ is a sheaf of abelian groups if $\mathscr{F}_x = \pi^{-1}(x)$ is an abelian group for every x and for any two sections φ,ψ of $\mathscr{F}$ over an open set U the map $U \ni x \to \varphi(x) - \psi(x)$ into $\mathscr{F}$ is a section. Similarly, we define the notion of sheaf of rings by requiring that all algebraic operations of the ring when applied to sections shall give sections. It is obvious that the sheaf $\mathscr{A}$ of germs of analytic functions on a complex manifold is a sheaf of rings. If $\mathscr{O}$ is a sheaf of rings on X, we define a sheaf of $\mathscr{O}$-modules $\mathscr{F}$ to be a sheaf of abelian groups such that $\mathscr{F}_x$ is an $\mathscr{O}_x$-module for every $x \in X$ and the product of a section of $\mathscr{O}$ and a section of $\mathscr{F}$ is a section of $\mathscr{F}$. When X is a complex manifold and $\mathscr{O}$ is the sheaf $\mathscr{A}$ of germs of analytic functions, we say that $\mathscr{F}$ is an *analytic sheaf.*

Example. If B is an analytic vector bundle over a complex manifold Ω, then the sheaf of germs of analytic sections of B is an analytic sheaf.

If $\mathscr{F}$ and $\mathscr{G}$ are sheaves of abelian groups on X, then a map $\varphi : \mathscr{F} \to \mathscr{G}$ is called a sheaf homomorphism if

(i) φ is continuous.

(ii) The restriction of φ to $\mathscr{F}_x$ is a homomorphism of the group $\mathscr{F}_x$ into $\mathscr{G}_x$.

The condition (i) means precisely that the composition of a section of $\mathscr{F}$ with φ is a section of $\mathscr{G}$.

The kernel $\ker \varphi$ and the image $\operatorname{im} \varphi$ of φ are *subsheaves* of $\mathscr{F}$ and $\mathscr{G}$, respectively; that is, these are open subsets of $\mathscr{F}$ and $\mathscr{G}$. This follows immediately from the definitions.

If $\mathscr{F}$ and $\mathscr{G}$ are sheaves of abelian groups on X and $\mathscr{F}$ is a subsheaf of $\mathscr{G}$, then $\mathscr{H}_x = \mathscr{G}_x/\mathscr{F}_x$ is an abelian group and $\mathscr{H} = \cup \mathscr{H}_x$ is a sheaf with the quotient topology; the sections of $\mathscr{H}$ are then locally (but not globally!) the images of sections of $\mathscr{G}$. If $\mathscr{F}$ and $\mathscr{G}$ are both analytic sheaves, then $\mathscr{G}/\mathscr{F}$ is an analytic sheaf.

Example 1. Let Ω be a complex manifold, let $\mathscr{M}_z$ be the quotient field of $\mathscr{A}_z$. Then $\mathscr{M} = \cup \mathscr{M}_z$ is a sheaf with the topology generated by the sets $\{f_z/g_z; z \in \omega\}$, where $\omega \subset \Omega$ is an open set, $f, g \in A(\omega)$ and $g_z \neq 0$ when $z \in \omega$. This sheaf is called the sheaf of germs of meromorphic functions; its sections are called meromorphic functions. (This definition is identical with Definition 1.4.1, although less explicit.) Note that in view of Theorem 6.2.3, part (ii), it is not possible to assign values in the extended complex plane to every germ of meromorphic function so that any meromorphic function in ω gives rise to a continuous function $\omega \to \mathbf{C} \cup \infty$.

To give the data of the first Cousin problem means precisely to give a section of the quotient sheaf $\mathscr{M}/\mathscr{A}$, and the Cousin problem thus consists in examining if a given section of $\mathscr{M}/\mathscr{A}$, is the image of a section of $\mathscr{M}$.

Example 2. If we remove from $\mathscr{M}$ the 0-section, we obtain a sheaf $\mathscr{M}^*$ of abelian groups (with multiplication as group operation). It contains as a sub-sheaf the sheaf $\mathscr{A}^*$ of invertible elements in $\mathscr{A}$, that is, the units in $\mathscr{A}$. The sheaf $\mathscr{D} = \mathscr{M}^*/\mathscr{A}^*$ is called the sheaf of divisors on Ω. The stalk $\mathscr{D}_z$ is an ordered abelian group if one defines the image of $\mathscr{A}_z - \{0\}$ in $\mathscr{D}_z$ to be the set of positive elements. From Theorem 6.2.2 it follows that $\mathscr{D}_z$ is the free abelian group generated by the classes in $\mathscr{D}_z$ of irreducible elements in A_z, and $\mathscr{D}_z$ has the natural order relation so $\mathscr{D}_z$ is a lattice. (Note that when $n = 1$ the sections of $\mathscr{D}$ are simply the integer valued functions on Ω which vanish except at a discrete set of points.) If φ and ψ are sections of $\mathscr{D}$, then $\sup(\varphi,\psi)$ and $\inf(\varphi,\psi)$ are also sections of $\mathscr{D}$ in the common domain of definition of φ and of ψ. It is sufficient to prove the statement about $\inf(\varphi,\psi)$ in a neighborhood of a point z_0 where $\inf(\varphi_{z_0},\psi_{z_0}) = 0$. For every z in a neighborhood ω of z_0, the germ $\varphi(z)$ is the image in $\mathscr{D}_z$ of the germ f_z of a function $f \in A(\omega)$, and $\psi(z)$ is the image of the germ g_z of a function $g \in A(\omega)$. The germs f_{z_0} and g_{z_0} are relatively prime, so it follows from Theorem 6.2.3, part (i), that f_z and g_z are relatively prime for all z in a neighborhood ω' of z_0. Hence $\inf(\varphi,\psi) = 0$ in ω', which proves the contention.

The second Cousin problem can be stated as follows: Given a section of the quotient sheaf $\mathscr{D} = \mathscr{M}^*/\mathscr{A}^*$, find a section of $\mathscr{M}^*$ which is mapped on the given section by the canonical map of $\mathscr{M}^*$ onto $\mathscr{M}^*/\mathscr{A}^*$.

Finally we define the direct sum $\mathscr{F} + \mathscr{G}$ of two sheaves of abelian groups $\mathscr{F}$ and $\mathscr{G}$; this is the sheaf whose stalk is the direct sum of $\mathscr{F}_x$ and $\mathscr{G}_x$, and the topology is induced by that of $\mathscr{F} \times \mathscr{G}$ if we consider $\mathscr{F} + \mathscr{G}$ as a subset of this topological product. This means that the sections of $\mathscr{F} + \mathscr{G}$ are the direct sums of the sections of $\mathscr{F}$ and those of $\mathscr{G}$. The sum of p copies of the sheaf $\mathscr{F}$ will be denoted by $\mathscr{F}^p$.

Arbitrary analytic sheaves are too general for our purposes, so we have to introduce a restrictive condition.

Definition 7.1.2. *An analytic sheaf $\mathscr{F}$ on the complex manifold Ω is said to be locally finitely generated if for every $z \in \Omega$ there exists a neighborhood $\omega \subset \Omega$ and a finite number of sections $f_1, \cdots, f_q \in \Gamma(\omega, \mathscr{F})$ so that $\mathscr{F}_z$ is generated by $(f_1)_z, \cdots, (f_q)_z$ as an A_z module for every $z \in \omega$.*

Example. The sheaf of germs of analytic sections of an analytic vector bundle is locally finitely generated.

Lemma 7.1.3. *If $\mathscr{F}$ is an analytic sheaf which is locally finitely generated and if $f_1, \cdots, f_q$ are sections of $\mathscr{F}$ in a neighborhood of z such that $(f_1)_z, \cdots, (f_q)_z$ generate $\mathscr{F}_z$, then $(f_1)_\zeta, \cdots, (f_q)_\zeta$ generate $\mathscr{F}_\zeta$ for every ζ in a neighborhood of z.*

Proof. Let $(g_1)_\zeta, \cdots, (g_r)_\zeta$ generate $\mathscr{F}$ for every ζ in a neighborhood of z. By hypothesis there exist analytic functions c_{ij} in a neighborhood of z such that

$$(g_i)_z = \sum_1^q (c_{ij})_z (f_j)_z, \qquad i = 1, \cdots, r.$$

Then we have

$$(g_i)_\zeta = \sum_1^q (c_{ij})_\zeta (f_j)_\zeta$$

for all ζ in a neighborhood of z, which proves the lemma.

If $f_1, \cdots, f_q \in \Gamma(\omega, \mathscr{F})$, where ω is an open subset of Ω, then the kernel of the sheaf homomorphism

$$\mathscr{A}^q \supset A_z^q \ni (g^1, \cdots, g^q) \to \sum_1^q g^j (f_j)_z \in \mathscr{F}_z \subset \mathscr{F}$$

is a subsheaf $\mathscr{R}(f_1, \cdots, f_q)$ of $\mathscr{A}^q$, over ω, called the sheaf of relations between $f_1, \cdots, f_q$.

Definition 7.1.4. *An analytic sheaf $\mathscr{F}$ on Ω is called coherent if*

(i) *$\mathscr{F}$ is locally finitely generated;*

(ii) *if ω is an open subset of Ω and $f_1, \cdots, f_q \in \Gamma(\omega, \mathscr{F})$, then the sheaf of relations $\mathscr{R}(f_1, \cdots, f_q)$ is locally finitely generated.*

Theorem 7.1.5. *Every locally finitely generated subsheaf of $\mathscr{A}^p$ is coherent.*

Proof. This is just another way of stating the Oka theorem (Theorem 6.4.1).

In particular, $\mathscr{A}^p$ is a coherent analytic sheaf, and so is the sheaf of germs of analytic sections of an analytic vector bundle.

Theorem 7.1.6. *If $\mathscr{F}$ is a coherent analytic sheaf on Ω and $f_1, \cdots, f_q$ are sections of $\mathscr{F}$ over the open subset ω of Ω, then the sheaf $\mathscr{R}(f_1, \cdots, f_q)$ is also coherent.*

Proof. This follows from Theorem 7.1.5 and part (ii) of Definition 7.1.4.

Example. There are subsheaves of $\mathscr{A}$ which are not coherent. For let ω be a subset of Ω with $\emptyset \neq \omega \neq \Omega$, set $\mathscr{F}_z = \mathscr{A}_z$ if $z \in \omega$ and $\mathscr{F}_z = 0$ if $z \in \Omega \setminus \omega$. This is a subsheaf of $\mathscr{A}$ if and only if ω is open. But a section of this sheaf over a connected open set which intersects $\Omega \setminus \omega$ must be 0, by the uniqueness of analytic continuation, so $\mathscr{F}$ is not finitely generated in any neighborhood of a boundary point of ω.

Theorem 7.1.7. *Let $\mathscr{G}$ be a subsheaf of the analytic sheaf $\mathscr{F}$. If two of the sheaves $\mathscr{F}, \mathscr{G}$ and $\mathscr{F}/\mathscr{G}$ are coherent, then all three are coherent.*

We leave the simple but rather lengthy proof as an exercise for the reader. (See also Serre [1] for a much more careful discussion of all topics touched upon in this section.)

7.2. Existence of global sections of a coherent analytic sheaf. In this paragraph we shall extend Corollary 5.6.3 to a theorem on existence of sections of an arbitrary coherent analytic sheaf. We first give a semi-global version. To abbreviate the statements, we shall say that sections $f_1, \cdots, f_q$ of an analytic sheaf $\mathscr{F}$ over an open set ω generate $\mathscr{F}$ there if the germs $(f_1)_z, \cdots, (f_q)_z$ generate the A_z-module $\mathscr{F}_z$ for every $z \in \omega$.

Theorem 7.2.1. *Let Ω be a Stein manifold, K an $A(\Omega)$-convex compact subset of Ω, and $\mathscr{F}$ a coherent analytic sheaf on a neighborhood of K. Then*

(i) *There exist finitely many sections $f_1, \cdots, f_q$ of $\mathscr{F}$ over a neighborhood of K which generate $\mathscr{F}$ there.*

(ii) *If $f_1, \cdots, f_q$ are sections of $\mathscr{F}$ over a neighborhood of K which generate $\mathscr{F}$ there and if f is an arbitrary section of $\mathscr{F}$ over a neighborhood of K, then one can find $c_1, \cdots, c_q$ analytic in a neighborhood of K so that $f = \sum_1^q c_j f_j$ there.*

In the proof we shall say that a compact $A(\Omega)$-convex set K has the property (E) if the statement of the theorem is true for every coherent analytic sheaf on a neighborhood of K. By the definition of a coherent analytic sheaf, every K consisting of one point only has the property (E). To prove the theorem we shall use an inductive procedure due to Cartan.

Lemma 7.2.2. *Let K be a compact $A(\Omega)$-convex subset of the Stein manifold Ω. Let $f \in A(\Omega)$ and assume that $K_a = \{z; z \in K, \operatorname{Re} f(z) = a\}$ has the property (E) for every real number a. Then K has the property (E).*

Proof. Let $\mathscr{F}$ be a coherent analytic sheaf on a neighborhood of K. Set

$$K_{a,b} = \{z; z \in K, a \leq \operatorname{Re} f(z) \leq b\}.$$

This is an $A(\Omega)$-convex set, since the condition $a \leq \operatorname{Re} f(z) \leq b$ can be written $|e^{f(z)}| \leq e^b$ and $|e^{-f(z)}| \leq e^{-a}$.

(i) To prove that there exists a finite number of sections of $\mathscr{F}$ over a neighborhood of K which generate $\mathscr{F}$ there, we introduce the supremum α of the set of all a such that $K_{-\infty,a}$ has this property. That set is not empty, for $K_{-\infty,a} = \emptyset$ if a is negative and $|a|$ is sufficiently large. We have to show that $\alpha = +\infty$. Thus assume that $\alpha < +\infty$. By the hypothesis in the theorem we can choose a neighborhood ω_1 of $K_{\alpha,\alpha} = K_\alpha$ where $\mathscr{F}$ is generated by a finite number of its sections $f_1, \cdots, f_p$. Choose $a < \alpha < b$ so that $\omega_1 \supset K_{a,b}$. By the definition of α we can then choose a neighborhood ω_2 of $K_{-\infty,a}$, where $\mathscr{F}$ is generated by a finite number of its sections $g_1, \cdots, g_q$. By hypothesis we can then find a neighborhood ω_3 of $K_{a,a}$ where there exists a matrix γ_1 with q lines and p columns and analytic coefficients such that

$$\gamma_1 f = g \quad \text{in } \omega_3;$$

here f is the column vector with components $f_1, \cdots, f_p$ and g is defined similarly. In fact, the existence of ω_3 follows from the validity of (ii) for $K_{a,a}$. In the same way we can find an analytic matrix γ_2 in a neighborhood ω_4 of $K_{a,a}$ with p lines and q columns, so that

$$\gamma_2 g = f \quad \text{in } \omega_4.$$

Shrinking the neighborhoods if necessary we can achieve that

$$\omega_1 \cap \omega_2 = \omega_3 = \omega_4.$$

Since the set $K_{-\infty,b}$ is $A(\Omega)$-convex and is contained in $\omega_1 \cup \omega_2$, it has a neighborhood ω contained in $\omega_1 \cup \omega_2$ which is an analytic polyhedron and therefore a Stein manifold (Lemma 5.3.7). If we replace the neighborhoods ω_j by their intersections with ω, then $\omega_1 \cup \omega_2 = \omega$ is a Stein manifold.

To obtain sections of $\mathscr{F}$ over ω we now wish to find analytic functions $u_1^j, \cdots, u_{p+q}^j$ in ω_j, $j = 1, 2$, so that the sections $\Sigma_1^p u_k^1 f_k$ and $\Sigma_1^q u_{p+k}^2 g_k$ of $\mathscr{F}$ over ω_1 and ω_2, respectively, coincide in $\omega_1 \cap \omega_2$. Writing $\{f,0\}$ and $\{0,g\}$ for the $(p+q)$-tuples (column vectors) $(f_1, \cdots, f_p, 0, \cdots, 0)$ and $(0, \cdots, 0, g_1, \cdots, g_q)$ of sections of $\mathscr{F}$, we can write this condition in the form

$$\langle \{f,0\}, u^1 \rangle = \langle \{0,g\}, u^2 \rangle \text{ in } \omega_1 \cap \omega_2. \tag{7.2.1}$$

Now we have

$$\{f,g\} = \begin{pmatrix} I_p & 0 \\ \gamma_1 & I_q \end{pmatrix} \{f,0\} = \begin{pmatrix} I_p & \gamma_2 \\ 0 & I_q \end{pmatrix} \{0,g\},$$

where I_p and I_q are the $p \times p$ and $q \times q$ unit matrices, respectively. It follows that

$$\{f,0\} = \begin{pmatrix} I_p & 0 \\ -\gamma_1 & I_q \end{pmatrix} \begin{pmatrix} I_p & \gamma_2 \\ 0 & I_q \end{pmatrix} \{0,g\}$$

and (7.2.1) is therefore fulfilled if $c_{21} u^1 = u^2$ in $\omega_1 \cap \omega_2$, where

$$c_{21} = c_{12}^{-1} = \begin{pmatrix} I_p & 0 \\ \gamma_2^{\,t} & I_q \end{pmatrix} \begin{pmatrix} I_p & -\gamma_1^{\,t} \\ 0 & I_q \end{pmatrix},$$

γ_j^t denoting the transpose of the matrix γ_j. Now consider the fiber bundle B over ω defined by the covering ω_1, ω_2 and the transition matrices c_{12} and c_{21}. According to Corollary 5.6.3 it has analytic sections over ω, and according to the construction of B every analytic section of B gives rise to a section of $\mathscr{F}$. Moreover, Corollary 5.6.3 implies that for every $z \in \omega$ the A_z-module generated by germs of analytic sections of B over ω is the module of all germs of local sections of B at z. Hence the A_z-module generated by the germs at z of sections of $\mathscr{F}$ over ω is equal to $\mathscr{F}_z$ for every $z \in \omega$. In view of Lemma 7.1.3 and the Borel–Lebesgue lemma, one can therefore find a finite number of sections of $\mathscr{F}$ over ω which A_z-generate $\mathscr{F}_z$ for every $z \in K_{-\infty,b}$. This is a contradiction with the definition of α which proves (i).

(ii) Let $f_1, \cdots, f_q$ be sections of $\mathscr{F}$ over a neighborhood of K which generate $\mathscr{F}$ there, and let f be an arbitrary section of $\mathscr{F}$ over a neighborhood of K. Introduce the supremum α of all a such that one can find

$c_1, \cdots, c_q$ analytic in a neighborhood of $K_{-\infty,a}$ so that $f = \Sigma_1^q c_j f_j$ there. We wish to show that $\alpha = +\infty$. Therefore, we assume that $\alpha < +\infty$ and aim at a contradiction as in part (i) of the proof. As there we can choose numbers a, b with $a < \alpha < b$, a neighborhood ω_1 of $K_{a,b}$, and a neighborhood ω_2 of $K_{-\infty,a}$, so that $\omega = \omega_1 \cup \omega_2$ is a Stein manifold and there exist analytic functions $c_1{}^k, \cdots, c_q{}^k$ in ω_k, $k = 1, 2$, for which

$$f = \sum_1^q c_j{}^k f_j \text{ in } \omega_k, \qquad k = 1, 2.$$

This implies that

$$\sum_1^q (c_j{}^1 - c_j{}^2) f_j = 0 \quad \text{in } \omega_1 \cap \omega_2.$$

Hence $(c_1{}^1 - c_1{}^2, \cdots, c_q{}^1 - c_q{}^2)$ is a section over $\omega_1 \cap \omega_2$ of the sheaf of relations $\mathscr{R}(f_1, \cdots, f_q)$, which is a coherent analytic sheaf in ω. By part (i) of the proof—which is valid for every coherent analytic sheaf—we can find a neighborhood ω' of K where this sheaf is generated by a finite number of its sections $r^1, \cdots, r^N$. In view of part (ii) of the hypothesis, it follows that

$$c_j{}^1 - c_j{}^2 = \sum_{k=1}^N \gamma_k r_j{}^k, \qquad j = 1, \cdots, q,$$

in a neighborhood of $K_{a,a}$, where γ_k are analytic functions. Shrinking ω_1 and ω_2 if necessary we may assume that this is true in $\omega_1 \cap \omega_2$ and that $\omega_1 \cup \omega_2 = \omega \subset \omega'$. Since the first Cousin problem in ω can be solved (Theorem 5.5.1), there exist functions $\gamma_k{}^\nu \in A(\omega_\nu)$, $k = 1, \cdots, N$, $\nu = 1, 2$, such that

$$\gamma_k = \gamma_k{}^2 - \gamma_k{}^1 \text{ in } \omega_1 \cap \omega_2.$$

Thus

$$c_j{}^1 + \sum_1^N \gamma_k{}^1 r_j{}^k = c_j{}^2 + \sum_1^N \gamma_k{}^2 r_j{}^k \quad \text{in } \omega_1 \cap \omega_2,$$

so the two sides of this equality define together a function $C_j \in A(\omega)$. Since r^k is a section of $\mathscr{R}(f_1, \cdots, f_q)$, it follows that

$$\sum_1^q C_j f_j = f \quad \text{in } \omega_1 \cup \omega_2 = \omega \supset K_{-\infty,b}.$$

This contradicts the definition of α, so it follows that $\alpha = +\infty$. The lemma is proved.

Proof of Theorem 7.2.1. We can choose a finite number of functions $F_1, \cdots, F_N \in A(\Omega)$ which form coordinate systems at every point in K. Then the set

$$\{z; z \in K, \operatorname{Re} F_j(z) = a_j, \operatorname{Re} i F_j(z) = b_j, j = 1, \cdots, N\}$$

has the property (E) for arbitrary a_j and b_j, since it is a discrete set of points. In view of Lemma 7.2.2, it follows that the set obtained by dropping one of the conditions $\operatorname{Re} F_j(z) = a_j$ and $\operatorname{Re} i F_j(z) = b_j$ still has the property (E). Repeating this argument $2N$ times, we conclude that K has the property (E).

Let Ω be a Stein manifold and $\mathscr{F}$ a coherent analytic sheaf on Ω. We shall now construct global sections of $\mathscr{F}$. First we introduce seminorms on sections of $\mathscr{F}$ over a neighborhood of a compact $A(\Omega)$-convex set K. To do so, we choose, using Theorem 7.2.1, a finite number of sections $f_1, \cdots, f_q$ of $\mathscr{F}$ over a neighborhood of K. Every section f of $\mathscr{F}$ over a neighborhood of K can then be written

$$f = \sum_1^q c_j f_j, \tag{7.2.2}$$

where the functions c_j are analytic in a neighborhood of K. We set

$$\|f\|_K = \inf_c \sup_{z \in K} \sum_1^q |c_j(z)|,$$

the infimum being taken over all c such that (7.2.2) is valid in an unspecified neighborhood of K. It is clear that $\|f\|_K$ may depend on the choice of the generators $f_1, \cdots, f_q$, but another choice must give an equivalent seminorm since passage from one system of generators to another is done by multiplication with a matrix whose entries are analytic in a neighborhood of K.

Lemma 7.2.3. *If f is a section of $\mathscr{F}$ over a neighborhood of K and if $\|f\|_K = 0$, then $f_z = 0$ for every z in the interior of K.*

Proof. Let f have the representation (7.2.2). If $\|f\|_K = 0$, we can for every $\varepsilon > 0$ find c_j^ε analytic in a neighborhood of K so that

$$f = \sum_1^q c_j^\varepsilon f_j \tag{7.2.2$'$}$$

in a neighborhood of K and $|c_j^\varepsilon(z)| < \varepsilon$ in K when $j = 1, \cdots, q$. By (7.2.2) and (7.2.2)$'$, we have

$$\gamma_z(c_1 - c_1^\varepsilon, c_2 - c_2^\varepsilon, \cdots, c_q - c_q^\varepsilon) \in \mathscr{R}_z(f_1, \cdots, f_q).$$

If z is in the interior of K, it follows from Theorem 2.2.3 that all derivatives of c_j^ε at z converge to 0 when $\varepsilon \to 0$, so Theorem 6.3.5 gives $\gamma_z(c_1, \cdots, c_q) \in \mathcal{R}_z(f_1, \cdots, f_q)$, which means that $f_z = 0$.

Lemma 7.2.4. *Let K and K' be $A(\Omega)$-convex compact sets with K contained in the interior of K'. If $g_1, g_2, \cdots$ is a sequence of sections of $\mathcal{F}$ over a neighborhood of K' such that*

$$\sum_1^\infty \|g_k\|_{K'} < \infty,$$

there exists a section g of $\mathcal{F}$ over a neighborhood of K such that

$$\|g - \sum_1^j g_k\|_K \to 0 \quad \text{when } j \to \infty.$$

Note that, by Lemma 7.2.3, g_z is uniquely determined by this condition when z is in the interior of K.

Proof. Let $f_1, \cdots, f_q$ be sections of $\mathcal{F}$ over a neighborhood of K' which generate $\mathcal{F}_z$ for every $z \in K'$. Then we can write

$$g_k = \sum_1^q c_{kj} f_j,$$

where c_{kj} are analytic functions in a neighborhood of K' and

$$\sum_{j=1}^q |c_{kj}(z)| \le \|g_k\|_{K'} + 2^{-k}, \qquad z \in K'.$$

This implies that the series

$$\sum_{k=1}^\infty c_{kj} = C_j$$

converges uniformly in K', so C_j is analytic in the interior of K'. Since $C_j - \Sigma_{k=1}^\nu c_{kj} \to 0$ uniformly in K when $\nu \to \infty$, the lemma follows if we set $g = \Sigma_1^q C_j f_j$. (We may assume that the norm $\| \ \|_K$ is defined by means of the same sections $f_1, \cdots, f_q$.)

Now let $K_1, K_2, \cdots$ be a sequence of compact $A(\Omega)$-convex subsets of Ω with $\cup_1^\infty K_p = \Omega$, each contained in the interior of the following one. From Lemma 7.2.4 we then obtain

Theorem 7.2.5. *Let g_p be a section of the coherent analytic sheaf $\mathcal{F}$ in Ω over a neighborhood of K_p and assume that for every fixed p*

$$\|g_i - g_j\|_{K_p} \to 0 \quad \text{when } i \text{ and } j \to \infty.$$

Then there is a section $g \in \Gamma(\Omega,\mathscr{F})$ *such that* $\|g - g_j\|_{K_p} \to 0, j \to \infty$, *for every* p.

Proof. Choose N_ν so that $\|g_i - g_j\|_{K_p} < 2^{-\nu}$ when i and j are $\geq N_\nu$ and $p \leq \nu$. Then the series with terms $g_{N_{\nu+1}} - g_{N_\nu}$ satisfies the hypotheses of Lemma 7.2.4 with K replaced by K_{p+1}. Hence there is a section h^p of $\mathscr{F}$ over a neighborhood of K_p such that $\|h^p - g_{N_\nu}\|_{K_p} \to 0$ when $\nu \to \infty$. We have $h^p = h^{p+1}$ in the interior of K_p, so there is a section $g \in \Gamma(\Omega,\mathscr{F})$ such that $g = h^p$ in the interior of K_p for every p. Hence $\|g - g_{N_\nu}\|_{K_{p-1}} \to 0$ for every p when $\nu \to \infty$. By the triangle inequality it follows that $\|g - g_j\|_{K_{p-1}} \to 0$ when $j \to \infty$, which completes the proof.

Corollary 7.2.6. $\Gamma(\Omega,\mathscr{F})$ *is a Fréchet space with the topology defined by the seminorms*

$$\Gamma(\Omega,\mathscr{F}) \ni f \to \|f\|_{K_p}, \qquad p = 1, 2, \cdots.$$

We can now extend the approximation theorems which we have proved for analytic functions or sections of analytic vector bundles.

Theorem 7.2.7. *Let K be a compact $A(\Omega)$-convex subset of the Stein manifold Ω, and let f be a section over a neighborhood of K of the coherent analytic sheaf $\mathscr{F}$ on Ω. Then there exists a sequence $f_j \in \Gamma(\Omega,\mathscr{F})$ such that $\|f - f_j\|_K \to 0$ when $j \to \infty$.*

Proof. We can choose a sequence of compact sets as above with $K_1 = K$. Put $f = g_1$. We shall for given $\varepsilon > 0$ and every p find a section g_p of $\mathscr{F}$ over a neighborhood of K_p such that

$$\|g_{p+1} - g_p\|_{K_\nu} < \varepsilon/2^p, \qquad p = 1, 2, \cdots; \nu \leq p.$$

By Theorem 7.2.5 this implies that there exists a section $g \in \Gamma(\Omega,\mathscr{F})$ such that $\|g - g_j\|_{K_p} \to 0$ when $j \to \infty$ for every p. In particular, we obtain $\|g - f\|_K = \lim \|g_j - g_1\|_{K_1} \leq \varepsilon$. If f_ν is chosen as this limit g when $\varepsilon = 1/\nu$, the theorem follows.

To show that sections g_p with the required properties exist, we assume that $g_1, \cdots, g_p$ have already been chosen. Let $h_1, \cdots, h_N$ be sections of $\mathscr{F}$ over a neighborhood of K_{p+1} which generate $\mathscr{F}$ there. We can write

$$g_p = \sum_1^N c_j h_j,$$

where c_j is analytic in a neighborhood of K_p. By Corollary 5.2.9 we can approximate the functions c_j arbitrarily closely on K_p by functions

$c_j' \in A(\Omega)$. Then

$$g_{p+1} = \sum_1^N c_j' h_j$$

is a section of $\mathscr{F}$ over a neighborhood of K_{p+1} which has the required properties if $\Sigma_1^N |c_j' - c_j|$ is small enough on K_p. The proof is complete.

Theorem 7.2.8. *Let Ω be a Stein manifold and $\mathscr{F}$ a coherent analytic sheaf on Ω. For every $z \in \Omega$ the A_z-module $\mathscr{F}_z$ is then generated by the germs at z of the sections $\in \Gamma(\Omega,\mathscr{F})$.*

This is known as Theorem A of Cartan.

Proof. Let $f_1, \cdots, f_N$ be sections of $\mathscr{F}$ over a neighborhood of z such that $(f_1)_z, \cdots, (f_N)_z$ generate $\mathscr{F}_z$. Since $\{z\}$ is $A(\Omega)$-convex, we can, by Theorem 7.2.7 for every $\varepsilon > 0$, choose sections $g_1, \cdots, g_N \in \Gamma(\Omega,\mathscr{F})$ so that

$$g_j - f_j = \sum_1^N c_{jk} f_k, \qquad j = 1, \cdots, N,$$

where c_{jk} are analytic in a neighborhood of z and $|c_{jk}(z)| < \varepsilon$ for all j and k. If ε is sufficiently small, the matrix $I + (c_{jk})$, where I is the $N \times N$ unit matrix, has an inverse (b_{jk}) which is analytic in a neighborhood of z. Since

$$f_j = \sum_1^N b_{jk} g_k$$

in a neighborhood of z, it follows that the germs of $g_1, \cdots, g_N$ at z are generators for $\mathscr{F}_z$.

We can also easily give a global form of part (ii) of Theorem 7.2.1:

Theorem 7.2.9. *Let Ω be a Stein manifold and $\mathscr{F}$ a coherent analytic sheaf on Ω. If $f, f_1, \cdots, f_q \in \Gamma(\Omega,\mathscr{F})$ and f_z is in the A_z-module generated by $(f_1)_z, \cdots, (f_q)_z$ for every $z \in \Omega$, one can find functions $c_1, \cdots, c_q \in A(\Omega)$ such that*

$$f = \sum_1^q c_j f_j.$$

Proof. Choose a sequence of compact $A(\Omega)$-convex sets K_p as in Theorem 7.2.5. The analytic subsheaf $\mathscr{F}'$ of $\mathscr{F}$ generated by the germs of the sections $f_1, \cdots, f_q$ is coherent, so by (ii) in Theorem 7.2.1 we can for every p choose $c_1{}^p, \cdots, c_q{}^p$ analytic in a neighborhood of K_p so that

$$f = \sum_{j=1}^q c_j^p f_j$$

there. Then

$$\sum_{j=1}^{q} (c_j^{p+1} - c_j^{p})f_j = 0 \text{ in a neighborhood of } K_p,$$

that is, $(c_1^{p+1} - c_1^{p}, \cdots, c_q^{p+1} - c_q^{p})$ is a section of the sheaf of relations $\mathscr{R}(f_1, \cdots, f_q)$ over a neighborhood of K_p. By Theorem 7.2.7 it can be approximated arbitrarily closely on K_p by sections of $\mathscr{R}(f_1, \cdots, f_q)$ over Ω, which means that the functions c_j^{p+1} can be changed by subtraction of such a section so that

$$\sup_{K_p} \sum_{j=1}^{q} |c_j^{p+1} - c_j^{p}| < 2^{-p}.$$

But this implies that $c_j = \lim_{p\to\infty} c_j^p$ exists for every j. We have $c_j \in A(\Omega)$, and from Lemma 7.2.3 it follows that $f = \Sigma_1^N c_j f_j$.

We shall now give applications of the last two theorems.

Theorem 7.2.10. *Let K be an $A(\Omega)$-convex compact subset of the Stein manifold Ω, and let B be the uniform closure on K of the restrictions to K of functions in $A(\Omega)$. Then the maximal ideal space of K can be identified with K, that is, every multiplicative linear functional on B is of the form $B \ni f \to f(z_0)$ for some $z_0 \in K$.*

Proof. Let I be an ideal in B. We have to show that if $I \neq B$, there is a point $z_0 \in K$ such that $f(z_0) = 0$ for every $f \in I$. Suppose that no such z_0 exists. For every $z \in K$ we can then find $f \in A(\Omega)$ with $f(z) \neq 0$. Hence we can choose a finite number of functions $f_1, \cdots, f_q \in A(\Omega)$ with no common zero in K. Let ω be a Stein manifold with $K \subset \omega \subset \Omega$ such that $\Sigma_1^N |f_j(z)| \neq 0$ when $z \in \omega$. Then the analytic subsheaf of $\mathscr{A}(\omega)$ generated by the sections $f_1, \cdots, f_N$ must be equal to $\mathscr{A}(\omega)$, for $(f_1)_z, \cdots, (f_N)_z$ generate A_z for every $z \in \omega$. By Theorem 7.2.9 we can therefore find $g_1, \cdots, g_q \in A(\omega)$ so that $\Sigma_1^q g_j f_j = 1$ in ω. But $g_1, \cdots, g_q$ can be uniformly approximated on K by functions in $A(\Omega)$, so their restrictions to K belong to B. Hence the identity belongs to I. This means that $I = B$, which completes the proof.

Remark. It is clear that it would have been sufficient to use Theorem 7.2.1 in the proof.

Note that Theorem 7.2.10 gives a simple proof of Theorem 5.4.3. In fact, Theorem 7.2.10 shows that every continuous multiplicative linear functional L on $A(\Omega)$ is of the form $L(f) = f(z_0)$ for some $z_0 \in \Omega$, if Ω is a Stein manifold. From the assumptions of Theorem 5.4.3 we therefore immediately obtain a map $\Omega_2 \to \Omega_1$.

Theorem 7.2.11. *Let Ω be a Stein manifold and V a submanifold (thus a closed set!). For every $z \notin V$ one can find $f \in A(\Omega)$ such that $f(z) \neq 0$, but $f = 0$ on V.*

Proof. Let I_z be the set of germs at z of analytic functions in a neighborhood of z which vanish on V. The sheaf $\mathcal{T} = \cup I_z$ is coherent. In fact, since $I_z = A_z$ if $z \notin V$, it is sufficient to show that $\mathcal{T}$ is finitely generated in a neighborhood of any point in V. We may assume that V is defined there by the equations $z_1 = \cdots = z_k = 0$, where $z_1, \cdots, z_n$ are local coordinates. By considering power series expansions of functions vanishing on V it is immediately seen that the germs of $z_1, \cdots, z_k$ at ζ generate I_ζ. Hence the theorem follows from Theorem 7.2.8.

7.3. Cohomology groups with values in a sheaf. The first Cousin problem, as stated in Theorem 5.5.1, leads naturally to the definition of cohomology groups with values in a sheaf. Let X be a topological space, $\mathcal{F}$ a sheaf of abelian groups on X, and $\mathcal{U} = \{U_i\}_{i \in I}$ an open covering of X. If p is a non-negative integer, we denote by $s = (s_0, \cdots, s_p)$ any element in I^{p+1}, and we set $U_s = U_{s_0} \cap \cdots \cap U_{s_p}$. A *p*-cochain c of the covering $\mathcal{U}$ with values in $\mathcal{F}$ is then a map which assigns to every $s \in I^{p+1}$ a section $c_s \in \Gamma(U_s, \mathcal{F})$ so that c_s is an alternating function of s (that is, c_s changes sign if two indices in s are permuted). Here we define $\Gamma(\emptyset, \mathcal{F}) = 0$, the abelian group with one element. The set $C^p(\mathcal{U}, \mathcal{F})$ of all *p*-cochains of $\mathcal{U}$ with values in $\mathcal{F}$ is of course an abelian group.

Imitating (5.5.1) and (5.5.2), we define a coboundary operator δ^p (which we often denote by δ only) from $C^p(\mathcal{U}, \mathcal{F})$ to $C^{p+1}(\mathcal{U}, \mathcal{F})$ as follows: If $c \in C^p(\mathcal{U}, \mathcal{F})$, then

$$(\delta^p c)_s = \sum_{j=0}^{p+1} (-1)^j c_{s_0 \cdots \hat{s}_j \cdots s_{p+1}},$$

where the notation $\hat{s}_j$ means that the index s_j shall be removed. With this definition, the data of the first Cousin problem consist of a 1-cochain c with values in $\mathcal{A}$, such that $\delta c = 0$, and the problem is to show that $c = \delta c'$ for some 0-cochain c'.

From the definition of δ it follows immediately that $\delta \circ \delta = 0$. If we introduce

$$Z^p(\mathcal{U}, \mathcal{F}) = \{c; c \in C^p(\mathcal{U}, \mathcal{F}), \delta c = 0\},$$

the group of *p-cocycles* with values in $\mathcal{F}$, and (with $C^{-1} = 0$)

$$B^p(\mathcal{U}, \mathcal{F}) = \{\delta c; c \in C^{p-1}(\mathcal{U}, \mathcal{F})\},$$

the group of *p-coboundaries*, then B^p is a subgroup of Z^p. We can therefore introduce the quotient group

$$H^p(\mathscr{U},\mathscr{F}) = Z^p(\mathscr{U},\mathscr{F})/B^p(\mathscr{U},\mathscr{F}),$$

which is called the pth *cohomology group* of $\mathscr{U}$ with values in $\mathscr{F}$. If c is a 0-cocycle, then $c_{s_0} - c_{s_1} = 0$ in $U_{s_0} \cap U_{s_1}$ for all s_0 and s_1, which means that there is a section $f \in \Gamma(X,\mathscr{F})$ with the restriction c_s to U_s for every s. Hence

$$H^0(\mathscr{U},\mathscr{F}) \approx \Gamma(X,\mathscr{F}).$$

Let us also note that Theorem 5.5.1 can be phrased as follows: $H^1(\mathscr{U},\mathscr{A}) = 0$ if X is a Stein manifold and $\mathscr{A}$ is the sheaf of germs of analytic functions on X. Our purpose is to extend this result to general coherent analytic sheaves and cohomology groups of higher order.

Now let $\mathscr{V} = \{V_j\}_{j\in J}$ be another covering of X, and assume that $\mathscr{V}$ is a refinement of $\mathscr{U}$, that is, that there exists a map $\rho: J \to I$ such that $V_j \subset U_{\rho(j)}$ for every $j \in J$. If $c \in C^p(\mathscr{U},\mathscr{F})$, we can then define a cochain $\rho c \in C^p(\mathscr{V},\mathscr{F})$ by setting $(\rho c)_s$ equal to the restriction of $c_{\rho(s_0)\cdots\rho(s_p)}$ to V_s. This map obviously commutes with the coboundary operators in $C^p(\mathscr{U},\mathscr{F})$ and $C^p(\mathscr{V},\mathscr{F})$, so it induces a map $\rho^*: H^p(\mathscr{U},\mathscr{F}) \to H^p(\mathscr{V},\mathscr{F})$.

Proposition 7.3.1. *The map ρ^* is independent of the choice of ρ.*

Proof. Let ρ' be another map with the same properties as ρ. We may assume that $p \geq 1$, since the statement is obvious when $p = 0$. After giving J a total ordering, we set, when $s_0 < s_1 < \cdots < s_p$,

$$(kc)_s = \sum_{j=0}^{p} (-1)^j c_{\rho(s_0)\cdots\rho(s_j)\rho'(s_j)\cdots\rho'(s_p)},$$

which defines a map of $C^{p+1}(\mathscr{U},\mathscr{F})$ into $C^p(\mathscr{V},\mathscr{F})$. A direct computation which we omit gives

$$(k\delta + \delta k)c = \rho' c - \rho c.$$

If c is a cocycle, it follows that $\rho' c - \rho c = \delta k c$ is a coboundary. Hence the maps ρ^* and ρ'^* are identical, which proves the proposition.

We have thus obtained a unique homomorphism $H^p(\mathscr{U},\mathscr{F}) \to H^p(\mathscr{V},\mathscr{F})$, which we denote by $\sigma(\mathscr{U},\mathscr{V})$. As a function of $\mathscr{U}$ and of $\mathscr{V}$, it has the obvious transitivity properties.

Proposition 7.3.2. *The homomorphism $\sigma(\mathscr{U},\mathscr{V})$ is injective when $p = 1$.*

Proof. Let $c \in Z^1(\mathscr{U},\mathscr{F})$ and assume that $\rho c = \delta\gamma$ for some $\gamma \in C^0(\mathscr{V},\mathscr{F})$.

This means explicitly that $c_{\rho(s)\rho(t)} = \gamma_t - \gamma_s$ in $V_s \cap V_t$ for all $s, t \in J$. Since c is a cocycle, it follows that

$$\gamma_t + c_{\rho(t)i} = \gamma_s + c_{\rho(s)i} \quad \text{in } U_i \cap V_s \cap V_t.$$

Hence there is a section $c_i \in \Gamma(U_i, \mathscr{F})$ such that $c_i = \gamma_s + c_{\rho(s)i}$ in $U_i \cap V_s$ for every s. In $U_i \cap U_j \cap V_s$ we obtain

$$c_j - c_i = \gamma_s + c_{\rho(s)j} - \gamma_s - c_{\rho(s)i} = c_{\rho(s)j} + c_{i\rho(s)} = c_{ij},$$

which proves that c is a coboundary. (Note that the proof is essentially a repetition of the second part of the proof of Theorem 5.5.2.)

If the coverings $\mathscr{U}$ and $\mathscr{V}$ are equivalent, that is, if $\mathscr{U}$ and $\mathscr{V}$ are refinements of each other, it is clear in view of Proposition 7.3.1 that the map $\sigma(\mathscr{U},\mathscr{V})$ is an isomorphism. The equivalence classes of open coverings form a set, since they can be identified with elements in the power set of the power set of X, and we have a natural partial ordering such that for any two (classes of) coverings there is a larger one (that is, finer one). We can therefore introduce the direct limit $H^p(X,\mathscr{F})$ of the groups $H^p(\mathscr{U},\mathscr{F})$ with the maps $\sigma(\mathscr{U},\mathscr{V})$. Explicitly, this means that the elements in $H^p(X,\mathscr{F})$ are the equivalence classes in the disjoint union of the groups $H^p(\mathscr{U},\mathscr{F})$, with an element in $H^p(\mathscr{U},\mathscr{F})$ and another in $H^p(\mathscr{V},\mathscr{F})$ identified if their images in $H^p(\mathscr{W},\mathscr{F})$ coincide for some refinement $\mathscr{W}$ of $\mathscr{U}$ and $\mathscr{V}$. In particular, if there exist arbitrarily fine coverings $\mathscr{U}$ with $H^p(\mathscr{U},\mathscr{F}) = 0$, then $H^p(X,\mathscr{F}) = 0$. When $p = 1$, it follows from Proposition 7.3.2 that we have a converse result: if $H^1(X,\mathscr{F}) = 0$, then $H^1(\mathscr{U},\mathscr{F}) = 0$ for every covering $\mathscr{U}$.

Proposition 7.3.3. *Let Ω be a C^∞ manifold which is countable at infinity, and $\mathscr{E}$ the sheaf of germs of C^∞ functions on Ω. If $\mathscr{F}$ is a sheaf of $\mathscr{E}$-modules on Ω, then $H^p(\mathscr{U},\mathscr{F}) = 0$ for every $p > 0$ and every covering $\mathscr{U}$. In particular, $H^p(\Omega,\mathscr{F}) = 0$ for every $p > 0$.*

Proof. Let $c \in Z^p(\mathscr{U},\mathscr{F})$, and let φ_ν be a partition of unity subordinate to the covering $\mathscr{U}$, that is,

(i) $\varphi_\nu \in C_0^\infty(U_{i_\nu})$ for a certain index i_ν.

(ii) All but a finite number of functions φ_ν vanish identically on any compact subset of Ω.

(iii) $$\sum \varphi_\nu = 1.$$

Then we can set, when $s \in I^p$,

$$c_s' = \sum \varphi_\nu c_{i_\nu s}.$$

It is clear that $c' \in C^{p-1}(\mathscr{U},\mathscr{F})$ and that

$$(\delta c')_s = \sum_{\nu} \sum_{j=0}^{p+1} \varphi_\nu (-1)^j c_{i_\nu s_0 \cdots \hat{s}_j \cdots s_p} = \sum_{\nu} \varphi_\nu c_s = c_s, \qquad s \in I^{p+1}.$$

Here we have of course used the fact that $\delta c = 0$. This completes the proof; we remark that it is essentially only a repetition of the first part of the proof of Theorem 1.4.5.

Let $\mathscr{F}$, $\mathscr{G}$, $\mathscr{H}$ be three sheaves of abelian groups on X and let φ, ψ be sheaf homomorphisms such that the sequence

$$0 \to \mathscr{F} \overset{\varphi}{\to} \mathscr{G} \overset{\psi}{\to} \mathscr{H} \to 0$$

is exact, that is, φ is injective, ψ is surjective, and the range of φ is equal to the kernel of ψ. It is clear that this gives rise to exact sequences

$$0 \to C^p(\mathscr{U},\mathscr{F}) \to C^p(\mathscr{U},\mathscr{G}) \to C^p(\mathscr{U},\mathscr{H}),$$

but the last map need not be surjective. We denote its image by $C_a{}^p(\mathscr{U},\mathscr{H})$ and call it the group of *liftable* cochains. By definition we then have an exact sequence

$$0 \to C^p(\mathscr{U},\mathscr{F}) \to C^p(\mathscr{U},\mathscr{G}) \to C_a{}^p(\mathscr{U},\mathscr{H}) \to 0.$$

Since δ maps $C_a{}^p(\mathscr{U},\mathscr{H})$ into $C_a{}^{p+1}(\mathscr{U},\mathscr{H})$, we can form the cohomology groups $H_a{}^p(\mathscr{U},\mathscr{H})$ of this complex; explicitly, $H_a{}^p(\mathscr{U},\mathscr{H}) = Z_a{}^p/B_a{}^p$ where $Z_a{}^b(B_a{}^b)$ is the group of all liftable p-cocycles (coboundaries of liftable $(p-1)$-cochains) with values in $\mathscr{H}$, belonging to the covering. Now consider the commutative diagram with exact columns:

$$\begin{array}{ccccc}
0 & & 0 & & 0 \\
\downarrow & & \downarrow & & \downarrow \\
C^{p-1}(\mathscr{U},\mathscr{F}) & \overset{\delta}{\to} & C^p(\mathscr{U},\mathscr{F}) & \overset{\delta}{\to} & C^{p+1}(\mathscr{U},\mathscr{F}) \\
\downarrow \varphi & & \downarrow \varphi & & \downarrow \varphi \\
C^{p-1}(\mathscr{U},\mathscr{G}) & \overset{\delta}{\to} & C^p(\mathscr{U},\mathscr{G}) & \overset{\delta}{\to} & C^{p+1}(\mathscr{U},\mathscr{G}) \\
\downarrow \psi & & \downarrow \psi & & \downarrow \psi \\
C_a{}^{p-1}(\mathscr{U},\mathscr{H}) & \overset{\delta}{\to} & C_a{}^p(\mathscr{U},\mathscr{H}) & \overset{\delta}{\to} & C_a{}^{p+1}(\mathscr{U},\mathscr{H}) \\
\downarrow & & \downarrow & & \downarrow \\
0 & & 0 & & 0
\end{array}$$

If $h \in Z_a{}^p(\mathscr{U},\mathscr{H})$, then $h = \psi g$ for some $g \in C^p(\mathscr{U},\mathscr{G})$, and $\psi\delta g = \delta\psi g = \delta h = 0$, that is, $\delta g \in \varphi C^{p+1}(\mathscr{U},\mathscr{F})$. Since φ is injective and $\delta\delta g = 0$, it follows that $\delta g \in \varphi Z^{p+1}(\mathscr{U},\mathscr{F})$. Now if $h \in B_a{}^p(\mathscr{U},\mathscr{H})$, we have $h = \delta h'$ for some $h' \in C_a{}^{p-1}(\mathscr{U},\mathscr{H})$, and we can find $g' \in C^{p-1}(\mathscr{U},\mathscr{G})$ with $\psi g' = h'$. Then $\delta g' - g \in \varphi C^p(\mathscr{U},\mathscr{F})$, which proves that $\delta g \in \varphi B^{p+1}(\mathscr{U},\mathscr{F})$. The cohomology class in $H^{p+1}(\mathcal{U},\mathcal{F})$ of $\varphi^{-1}\delta g$ is therefore uniquely defined by the cohomology class of h in $H_a^p(\mathcal{U},\mathcal{H})$, so we obtain a homomorphism

$$\delta^*: H_a^p(\mathcal{U},\mathcal{H}) \to H^{p+1}(\mathcal{U},\mathcal{F}).$$

Theorem 7.3.4. *The sequence*

$$0 \to H^0(\mathcal{U},\mathcal{F}) \xrightarrow{\varphi^*} H^0(\mathcal{U},\mathcal{G}) \xrightarrow{\psi^*} H_a^0(\mathcal{U},\mathcal{H}) \xrightarrow{\delta^*} H^1(\mathcal{U},\mathcal{F}) \xrightarrow{\varphi^*}$$

$$\to H^1(\mathcal{U},\mathcal{G}) \xrightarrow{\psi^*} H_a^1(\mathcal{U},\mathcal{H}) \xrightarrow{\delta^*} H^2(\mathcal{U},\mathcal{F}) \to \cdots$$

is exact.

Here the maps φ^* and ψ^* are obtained from φ and ψ in the obvious way, using the fact that the maps of cochains defined by φ and ψ commute with the coboundary operators.

Proof of Theorem 7.3.4. The exactness at $H^p(\mathcal{U},\mathcal{G})$ is obvious. Next we prove the exactness at $H_a^p(\mathcal{U},\mathcal{H})$. Let $h \in Z_a^p(\mathcal{U},\mathcal{H})$. From the diagram above we then see that h is the image of a cocycle $g \in Z^p(\mathcal{U},\mathcal{G})$, if and only if the cohomology class of h is in the kernel of δ^*. To prove exactness at $H^p(\mathcal{U},\mathcal{F})$, we note that the range of δ^* in $H^p(\mathcal{U},\mathcal{F})$ consists of the cohomology classes of all cocycles in $Z^p(\mathcal{U},\mathcal{F}) \cap \varphi^{-1}B^p(\mathcal{U},\mathcal{G})$, that is, precisely the cohomology classes which are mapped to 0 by φ^*.

If $\mathcal{V} = \{V_j\}_{j\in J}$ is a refinement of $\mathcal{U}$ and $\rho: J \to I$ is a map such that $V_j \subset U_{\rho(j)}$ for every $j \in J$, we can as above define a map

$$\rho^*: H_a^p(\mathcal{U},\mathcal{H}) \to H_a^p(\mathcal{V},\mathcal{H}),$$

and the proof of Proposition 7.3.1 applies without change to show that ρ^* is independent of the choice of ρ; we denote this homomorphism also by $\sigma(\mathcal{U},\mathcal{V})$. Again we have the obvious transitivity properties.

Proposition 7.3.5. *If X is paracompact (that is, if X satisfies the Hausdorff separation axiom and every covering of X has a locally finite refinement), then the natural map of the direct limit of $H_a^p(\mathcal{U},\mathcal{H})$ into $H^p(X,\mathcal{H})$ is an isomorphism onto.*

Proof. The statement is an immediate consequence of the following lemma:

Lemma 7.3.6. *If $h \in C^p(\mathcal{U},\mathcal{H})$, one can find a refinement $\mathcal{V} = \{V_j\}_{j\in J}$ and a map $\rho: J \to I$, such that $\rho h \in C_a^p(\mathcal{V},\mathcal{H})$.*

Proof. Since X is paracompact we may assume that $\mathcal{U}$ is locally finite, and since X is normal we can choose an open covering $\{W_i\}_{i\in I}$

such that $\overline{W}_i \subset U_i$. For every $x \in X$ we now choose an open neighborhood V_x of x such that

(i) If $x \in U_s (s \in I^{p+1})$, then $V_x \subset U_s$ and there is a section of $\mathscr{G}$ over V_x which is mapped on h_s by ψ.

(ii) If $x \in W_i$, it follows that $V_x \subset W_i$, and if $V_x \cap W_i \neq \varnothing$, it follows that $V_x \subset U_i$.

Such a neighborhood V_x exists since, by the definition of sheaf homomorphism, it is possible to lift every section of $\mathscr{H}$ to a section of $\mathscr{G}$ locally, and there are only a finite number of additional conditions on V_x to satisfy. Define $\rho : X \to I$ so that $x \in W_{\rho(x)}$, hence $V_x \subset W_{\rho(x)}$. If $V_{x_0} \cap \cdots \cap V_{x_p} \neq \varnothing$, it follows that V_{x_0} meets $W_{\rho(x_k)}$ for every k, hence $V_{x_0} \subset U_{\rho(x_0)} \cap \cdots \cap U_{\rho(x_p)}$ in view of (ii). The section $h_{\rho(x_0)\ldots\rho(x_p)}$ of $\mathscr{H}$ can therefore over $V_{x_0} \cap \cdots \cap V_{x_p}$ be lifted to a section of $\mathscr{G}$, which means that ρh is a liftable cochain in the covering $\{V_x\}_{x \in X}$.

From Proposition 7.3.5 and Theorem 7.3.4 we obtain by an obvious argument, which we leave as an exercise,

Theorem 7.3.7. *If X is paracompact and $0 \to \mathscr{F} \to \mathscr{G} \to \mathscr{H} \to 0$ is an exact sequence of sheaves of abelian groups on X, then the sequence*

$$0 \to H^0(X,\mathscr{F}) \to H^0(X,\mathscr{G}) \to H^0(X,\mathscr{H}) \to H^1(X,\mathscr{F})$$
$$\to H^1(X,\mathscr{G}) \to H^1(X,\mathscr{H}) \to H^2(X,\mathscr{F}) \to \cdots$$

is exact.

7.4. The cohomology groups of a Stein manifold with coefficients in a coherent analytic sheaf. Using the existence theorems for the $\bar{\partial}$ operator, we have already proved in Theorem 5.5.1 that $H^1(\Omega,\mathscr{A}) = 0$ if Ω is a Stein manifold and $\mathscr{A}$ the sheaf of germs of analytic functions. We shall now prove that existence theorems for the $\bar{\partial}$ operator are completely equivalent to statements involving $H^p(\Omega,\mathscr{A})$ (the Dolbeault isomorphism):

Theorem 7.4.1. *Let Ω be a complex manifold which is countable at infinity and let $\mathscr{U} = \{U_i\}_{i \in I}$ be a covering where each U_i is a Stein manifold. Then $H^p(\mathscr{U},\mathscr{A})$ is for $p > 0$ isomorphic to the quotient space*

$$\{f; f \in C^\infty_{(0,p)}(\Omega),\ \bar{\partial} f = 0\}/\{\bar{\partial} g;\ g \in C^\infty_{(0,p-1)}(\Omega)\},$$

and $H^p(\Omega,\mathscr{A})$ is isomorphic to $H^p(\mathscr{U},\mathscr{A})$.

Proof. Denote by $\mathscr{E}_q$ the sheaf of germs of C^∞ forms of type $(0,q)$ and consider the exact sequence of sheaf homomorphisms

$$0 \to \mathscr{Z}_q \xrightarrow{i} \mathscr{E}_q \xrightarrow{\bar{\partial}} \mathscr{Z}_{q+1} \to 0,$$

where $\mathscr{Z}_q$ is the sheaf of germs of $\bar{\partial}$ closed forms of type $(0,q)$. The exactness follows from Theorem 2.3.3. From Corollary 5.2.6 we even obtain that the sequence

$$0 \to C^p(\mathscr{U},\mathscr{Z}_q) \to C^p(\mathscr{U},\mathscr{E}_q) \to C^p(\mathscr{U},\mathscr{Z}_{q+1}) \to 0$$

is exact, for the intersection of any number of the sets U_i is also a Stein manifold, since it is obviously holomorph-convex. In other words, we have $C_a^p(\mathscr{U},\mathscr{Z}_{q+1}) = C^p(\mathscr{U},\mathscr{Z}_{q+1})$, so Theorem 7.3.4 gives the exactness of the sequence

$$\begin{aligned} 0 &\to \Gamma(\Omega,\mathscr{Z}_q) \to \Gamma(\Omega,\mathscr{E}_q) \xrightarrow{\bar{\partial}} \Gamma(\Omega,\mathscr{Z}_{q+1}) \to H^1(\mathscr{U},\mathscr{Z}_q) \\ &\to H^1(\mathscr{U},\mathscr{E}_q) \to H^1(\mathscr{U},\mathscr{Z}_{q+1}) \to H^2(\mathscr{U},\mathscr{Z}_q) \to H^2(\mathscr{U},\mathscr{E}_q) \to \cdots. \end{aligned}$$

Using Proposition 7.3.3, we conclude that

$$H^p(\mathscr{U},\mathscr{Z}_{q+1}) \approx H^{p+1}(\mathscr{U},\mathscr{Z}_q), \qquad p \geq 1,$$

$$\Gamma(\Omega,\mathscr{Z}_{q+1})/\bar{\partial}\Gamma(\Omega,\mathscr{E}_q) \approx H^1(\mathscr{U},\mathscr{Z}_q).$$

Hence we obtain, when $p > 0$,

$$\begin{aligned} H^p(\mathscr{U},\mathscr{A}) = H^p(\mathscr{U},\mathscr{Z}_0) &\approx H^{p-1}(\mathscr{U},\mathscr{Z}_1) \approx \cdots \approx H^1(\mathscr{U},\mathscr{Z}_{p-1}) \\ &\approx \Gamma(\Omega,\mathscr{Z}_p)/\bar{\partial}\Gamma(\Omega,\mathscr{E}_{p-1}), \end{aligned}$$

which proves the theorem if we note that there exist arbitrarily fine "Stein coverings."

Corollary 7.4.2. *If Ω is a Stein manifold, then $H^p(\Omega,\mathscr{A}) = 0$ for every $p > 0$. More precisely: $H^p(\mathscr{U},\mathscr{A}) = 0$ for every covering $\mathscr{U} = \{U_i\}_{i\in I}$, where each U_i is a Stein manifold. (When $p = 1$, this condition can be dropped by Proposition 7.3.2.)*

We shall now extend the result just proved to general coherent analytic sheaves. This is known as the Cartan theorem B.

Theorem 7.4.3. *If $\mathscr{F}$ is a coherent analytic sheaf on the Stein manifold Ω, then $H^p(\Omega,\mathscr{F}) = 0$ for every $p > 0$.*

Proof. Let $\mathscr{U} = \{U_i\}_{i\in I}$ be a covering where each U_i is a Stein manifold which is relatively compact in Ω. Let the Stein manifold Ω' be relatively compact in Ω and set $U_i' = U_i \cap \Omega'$. We shall first prove that

$H^p(\mathcal{U}',\mathcal{F}) = 0$ when $p > 0$ for every coherent analytic sheaf $\mathcal{F}$ on Ω. To do so, we choose $f_1, \cdots, f_q \in \Gamma(\Omega,\mathcal{F})$ which generate $\mathcal{F}$ in Ω'. This is possible by Theorem 7.2.8, Lemma 7.1.3, and the Borel–Lebesgue lemma. We then obtain an exact sequence of sheaf homomorphisms

$$0 \to \mathcal{R}(f_1, \cdots, f_q) \to \mathcal{A}^q \to \mathcal{F} \to 0,$$

where all the sheaves occurring are to be considered as sheaves on Ω'. By Theorem 7.2.9 and the fact that each intersection of the sets U_i' is a Stein manifold, we even have an exact sequence

$$0 \to C^p(\mathcal{U}',\mathcal{R}(f_1, \cdots, f_q)) \to C^p(\mathcal{U}',\mathcal{A}^q) \to C^p(\mathcal{U}',\mathcal{F}) \to 0,$$

that is, all cochains of $C^p(\mathcal{U}',\mathcal{F})$ are liftable. Hence we obtain an exact sequence

$$H^p(\mathcal{U}',\mathcal{A}^q) \to H^p(\mathcal{U}',\mathcal{F}) \to H^{p+1}(\mathcal{U}',\mathcal{R}) \to H^{p+1}(\mathcal{U}',\mathcal{A}^q).$$

Since the groups to the right and to the left are 0 by Corollary 7.4.2, we obtain $H^p(\mathcal{U}',\mathcal{F}) \approx H^{p+1}(\mathcal{U}',\mathcal{R})$. If it has already been proved that $H^{p+1}(\mathcal{U}',\mathcal{G}) = 0$ for every coherent analytic sheaf $\mathcal{G}$ on Ω, it follows that $H^p(\mathcal{U}', \mathcal{F}) = 0$ for every coherent analytic sheaf on Ω. But if the covering is chosen so that more than N sets U_i always have an empty intersection, which is possible if $N > 2n$, it is obvious that $H^p(\mathcal{U}',\mathcal{F}) = 0$ when $p > N$ for every sheaf $\mathcal{F}$, which completes the proof of the fact that $H^p(\mathcal{U}',\mathcal{F}) = 0$ when $p > 0$.

To pass from $\mathcal{U}'$ to $\mathcal{U}$ we have to distinguish two cases. (Cf. the proof of Theorem 2.7.8.)

(a) $p > 1$. Choose an increasing sequence of Stein manifolds $\Omega^1, \Omega^2, \cdots$ which are relatively compact in Ω and whose union is equal to Ω. Let c be a cocycle in $C^p(\mathcal{U},\mathcal{F})$ and let c^j be its restriction to a cocycle in $C^p(\mathcal{U}^j,\mathcal{F})$, where $\mathcal{U}^j = \{\Omega^j \cap U_i\}_{i\in I}$. We have then found that there is a cochain $b^j \in C^{p-1}(\mathcal{U}^j,\mathcal{F})$ such that $\delta b^j = c^j$. The difference between b^j and the restriction of b^{j+1} to Ω^j is thus a cocycle, hence equal to δa for some $a \in C^{p-2}(\mathcal{U}^j,\mathcal{F})$. Let $a' \in C^{p-2}(\mathcal{U},\mathcal{F})$ be defined so that $a_s' = a_s$ when $U_s \subset \Omega^j$ and $a_s' = 0$ otherwise. By subtracting from b^{j+1} the restriction to Ω^{j+1} of the coboundary of a', we attain that $(b^{j+1})_s = (b^j)_s$ if $U_{s_k} \subset \Omega^j$ for every k. Since each U_i is relatively compact in Ω by assumption, this means that $(b^j)_s$ is for every s independent of j when j is large, so b^j converges in an obvious sense when $j \to \infty$ to a cochain $b \in C^{p-1}(\mathcal{U},\mathcal{F})$ with $\delta b = c$.

(b) Now assume that $p = 1$. If we choose the cochains b^j as above, then the restriction of b^{j+1} to Ω^j differs from b^j by a 0-cocycle, that is, a section of $\mathscr{F}$ over Ω^j. Let K^j be a compact $A(\Omega)$-convex subset of Ω^j, chosen so that K^j is in the interior of K^{j+1} for every j and $\cup K^j = \Omega$. By Theorem 7.2.7 we can then find a section $a \in \Gamma(\Omega,\mathscr{F})$ such that

$$\|b^{j+1} - b^j - a\|_{K^\nu} < 2^{-j}, \qquad \nu \leq j.$$

We subtract from b^{j+1} the cocycle in $C^0(\mathscr{U}^{j+1},\mathscr{F})$ defined by a, and thus attain successively that for every j

$$\|b^{j+1} - b^j\|_{K^\nu} < 2^{-j}, \qquad \nu \leq j,$$

where of course $b^{j+1} - b^j$ denotes the section $\in \Gamma(\Omega^j,\mathscr{F})$ corresponding to the difference between the restriction of b^{j+1} to Ω^j and b^j. But then it follows from Lemma 7.2.4 that $b^k - b^j$ converges to a section s^j of $\mathscr{F}$ over the interior of K^j when $k \to \infty$ for fixed j. It is clear that $b^j + s^j = b^{j-1} + s^{j-1}$ in the interior of K^{j-1}, so there is a well defined 0-cochain $b \in C^0(\mathscr{U},\mathscr{F})$ such that the restriction of b to the interior of K^j is $b^j + s^j$ for every j. Since $\delta b^j = c$ in Ω^j for every j, we obtain $\delta b = c$, which completes the proof of the fact that $H^p(\mathscr{U},\mathscr{F}) = 0$, for every $p > 0$. From the fact that there exist arbitrarily fine coverings $\mathscr{U}$ with the properties we have required, it now follows that $H^p(\Omega,\mathscr{F}) = 0$. The proof is complete.

In the applications we shall repeatedly use the fact that if

$$0 \to \mathscr{F} \to \mathscr{G} \to \mathscr{H} \to 0$$

is an exact sequence of sheaf homomorphisms on X and if $H^1(X,\mathscr{F}) = 0$, then the map $\Gamma(X,\mathscr{G}) \to \Gamma(X,\mathscr{H})$ is onto. (Cf. Theorems 7.3.4 and 7.3.7.) This gives immediately a new proof of Theorem 7.2.9 if we consider the sheaf homomorphism

$$0 \to \mathscr{R}(f_1,\cdots,f_q) \to \mathscr{A}^q \to \mathscr{F} \to 0.$$

If with the notations of Example 1 in section 7.1 we consider the exact sequence

$$0 \to \mathscr{A} \to \mathscr{M} \to \mathscr{M}/\mathscr{A} \to 0,$$

it follows that the map $\Gamma(\Omega,\mathscr{M}) \to \Gamma(\Omega,\mathscr{M}/\mathscr{A})$ is onto if Ω is a Stein manifold, so we have again found that the first Cousin problem can be solved then. To study the second Cousin problem we consider the exact sequence

$$0 \to \mathscr{A}^* \to \mathscr{M}^* \to \mathscr{D} \to 0$$

with the notations of Example 2 of section 7.1. (The sheaves $\mathscr{A}^*$ and

$\mathscr{M}^*$ are multiplicative groups.) Then we obtain the exact sequence

$$\Gamma(\Omega,\mathscr{M}^*) \to \Gamma(\Omega,\mathscr{D}) \to H^1(\Omega,\mathscr{A}^*)$$

and conclude that the second Cousin problem is solvable for a section of $\mathscr{D}$ if and only if its image in $H^1(\Omega,\mathscr{A}^*)$ is 0. To study this group we consider the exact sequence

$$0 \to \mathbf{Z} \to \mathscr{A} \xrightarrow{\exp(2\pi i\,\cdot\,)} \mathscr{A}^* \to 0,$$

where $\mathbf{Z}$ is the constant sheaf of the additive group of integers and the second map is the map $f \to \exp(2\pi i f)$. The exact cohomology sequence gives the exact sequence

$$H^1(\Omega,\mathscr{A}) \to H^1(\Omega,\mathscr{A}^*) \to H^2(\Omega,\mathbf{Z}) \to H^2(\Omega,\mathscr{A}),$$

where the groups to the left and right are both 0. Hence

$$H^1(\Omega,\mathscr{A}^*) \approx H^2(\Omega,\mathbf{Z}).$$

The map $\Gamma(\Omega,\mathscr{D}) \to H^1(\Omega,\mathscr{A}^*)$ can therefore be regarded as a map

$$\Gamma(\Omega,\mathscr{D}) \to H^2(\Omega,\mathbf{Z}), \tag{7.4.1}$$

and the second Cousin problem is solvable precisely for the divisors belonging to the kernel of this map. The map (7.4.1) is onto. Indeed, Theorem 5.5.3 means that every 1-cocycle of $\mathscr{A}^*$ is the image of a positive divisor, that is, a section of $\mathscr{D}$, which is at every point the divisor of an analytic function. Hence even the positive divisors are mapped onto $H^1(\Omega,\mathscr{A}^*)$, so we obtain

Theorem 7.4.4. *The second Cousin problem on a Stein manifold can be solved for an arbitrary divisor if and only if* $H^2(\Omega,\mathbf{Z}) = 0$.

Note that the discussion we have just made is only a less explicit form of that given in section 5.5.

Theorem 7.4.5. *If F is a meromorphic function on a Stein manifold Ω with $H^2(\Omega,\mathbf{Z}) = 0$, then one can find analytic f and g so that $F = f/g$ and the germs f_z, g_z are relatively prime at every point; this representation is unique apart from an analytic nonvanishing factor.*

Proof. F defines a section $\operatorname{div} F$ of $\mathscr{D}$, and we proved in section 7.1 that $\sup(0,\operatorname{div} F) = d^+$ and $\sup(0,-\operatorname{div} F) = d^-$ are sections of $\mathscr{D}$ and that $\operatorname{div} F = d^+ - d^-$. By Theorem 7.4.4 we can therefore find meromorphic functions f^+ and f^- with these divisors. But f^+ and f^- are then analytic, and f^+/f^- defines the same divisor as F. Hence $f^+/f^- = F/h$ for some analytic nonvanishing h and, if we set $f = hf^+, g = f^-$, the theorem is proved since the uniqueness is obvious.

Without making any hypothesis concerning $H^2(\Omega, \mathbf{Z})$, we obtain a weaker result.

Theorem 7.4.6. *Every meromorphic function F in the Stein manifold Ω can be written in the form f/g where f and $g \in A(\Omega)$.*

Proof. Let $\mathscr{F}_z$ be the ideal of germs of analytic functions g_z at z such that $g_z F_z$ is the germ of an analytic function. Since every point in Ω has a neighborhood where Theorem 7.4.5 is applicable, we can there obtain an analytic function g such that $\mathscr{F}_z$ is equal to $g_z A_z$ for every z in the neighborhood. Hence $\mathscr{F}$ is a coherent analytic subsheaf of $\mathscr{A}$. Let g be a nontrivial section of this sheaf, which exists by Theorem 7.2.8. Then $gF = f$ is analytic, which proves the theorem. Note that by Theorem 7.2.8 we can choose g as a generator of $\mathscr{F}$ at any given point, that is, so that f_z and g_z are relatively prime at a given point, but we cannot achieve that this is true everywhere in Ω.

7.5. The de Rham theorem. Let Ω be a C^∞ manifold which is countable at infinity and denote by $\mathscr{E}_q$ the sheaf of germs of C^∞ q-forms on Ω. Let $\mathscr{Z}_q$ be the sheaf of germs of such forms which are annihilated by the exterior differential operator d. Then we have an exact sequence of sheaf homomorphisms

$$0 \to \mathscr{Z}_q \to \mathscr{E}_q \xrightarrow{d} \mathscr{Z}_{q+1} \to 0.$$

In fact, if f is a $(q+1)$-form with $df = 0$ in an interval in $\mathbf{R}^n$, defined by $a_j < x_j < b_j, j = 1, \cdots, n$, one can find a q-form u with $du = f$. This fact, the Poincaré lemma, follows by the argument used to prove Theorem 2.3.3 if the existence of solutions of the Cauchy–Riemann equation in one complex variable is replaced by the existence of a primitive function for every C^∞ function of one real variable. The proof of Theorem 7.4.1 can now be applied and gives, since $\mathscr{Z}_0$ is the constant sheaf $\mathbf{C}$:

Theorem 7.5.1. *The cohomology group $H^p(\Omega, \mathbf{C})$ is for every $p > 0$ isomorphic to the quotient space*

$$\{f; f \in \Gamma(\Omega, \mathscr{E}_p), df = 0\}/\{dg; g \in \Gamma(\Omega, \mathscr{E}_{p-1})\}.$$

This completes the proof of Theorem 2.7.10. To complete the proof of Theorem 2.7.11 we need to use special coverings. It is therefore convenient now to consider only open subsets Ω of $\mathbf{R}^n$, although the arguments can be applied in general in view of a theorem of Weil [2]. Thus, choose a covering $\mathscr{U} = \{U_i\}_{i\in I}$ of Ω where each U_i is an open interval contained in Ω and I is countable. Then the sequence

$$0 \to C^p(\mathscr{U},\mathscr{Z}_q) \to C^p(\mathscr{U},\mathscr{E}_q) \xrightarrow{d} C^p(\mathscr{U},\mathscr{Z}_{q+1}) \to 0$$

is also exact, and we obtain by the arguments used in proving Theorem 7.4.1

$$H^p(\mathscr{U},\mathbf{C}) \approx \Gamma(\mathscr{Z}_p)/d\mathscr{E}_{p-1}.$$

If we fix for each p an operator giving an inverse of the operator d in the above sequence, it is easily seen that the proof assigns to each $f \in \Gamma(\mathscr{Z}_p)$ a cocycle $c \in H^p(\mathscr{U},\mathbf{C})$ such that c_s depends continuously on f (in the C^∞ topology) for every $s \in I^{p+1}$, and $f \in d\mathscr{E}_{p-1}$ if and only if c is a coboundary. Since the equation $\delta b = c$ is a linear system of equations of the kind discussed in Lemma 6.3.7, it follows that $d\mathscr{E}_{p-1}$ is closed in $\Gamma(\mathscr{Z}_p)$. This completes the proof of Theorem 2.7.11.

Finally, we shall prove Theorem 2.7.12. It is enough to consider finite coverings of K of the form $\{U_i \cap K\}_{i\in I}$ where U_i is an open set in $\mathbf{C}^n$, and by shrinking U_i if necessary we can attain that either $U_s \cap K \neq \emptyset$ or $\bar{U}_s \cap K = \emptyset$ for every $s \in I^{p+1}$ and any integer p. Let Ω' be a Runge domain contained in $\cup U_i$ such that $\Omega' \cap U_s \neq \emptyset$ implies $K \cap U_s \neq \emptyset$. The existence of Ω' follows from the fact that K has a fundamental family of neighborhoods which are Runge domains and the condition on the covering made initially. Now let c be a cocycle $\in C^r(\{U_i \cap K\}, \mathbf{C})$. We may assume that c_s has a constant value in $U_s \cap K$ for every $s \in I^{r+1}$, for this could be achieved by passage to a finer covering since the value is locally constant. Then there is a unique cochain $c' \in C^r(\{U_i \cap \Omega'\}, \mathbf{C})$ which for every s has a constant value in $U_s \cap \Omega'$, equal to the value of c in $U_s \cap K$. It is clear that c' is a cocycle, so if $r \geq n$ it follows from Theorem 2.7.11 that for some refinement of the covering $\{U_i \cap \Omega'\}$ of Ω' the cocycle c' is mapped to a coboundary. If we restrict this covering to K, we conclude that c is mapped to a coboundary in this covering, hence $H^r(K,\mathbf{C}) = 0$ if $r \geq n$. The proof is complete.

7.6. Cohomology with bounds and constant coefficient differential equations. We proved Theorem B (Theorem 7.4.3) by combining the existence theory for the $\bar{\partial}$ operator with the local results obtained in Chapter VI. Now we have given precise bounds for the solutions of the $\bar{\partial}$ equation in $\mathbf{C}^n$ (section 4.4). We shall prove in this section that combined with a more precise discussion of the local situation they lead to a

"Theorem B with bounds." This result contains existence theorems for overdetermined systems of constant coefficient differential operators.

The first step is to give a quantitative version of Theorem 7.4.1 In doing so we shall use some standard coverings of $\mathbf{C}^n$, which will also be useful in the rest of the section. If ν is a non-negative integer, we denote by $\mathscr{U}^{(\nu)}$ the covering of $\mathbf{C}^n$ which consists of the cubes $U_g{}^{(\nu)}$ with side equal to $2\ 3^{-\nu}$ and center at $g\ 3^{-\nu}$, where g runs through the set I of points in $\mathbf{C}^n$ with integral coordinates. (The edges are required to be parallel to the coordinate axes.) For every ν and g we can then find precisely one g' such that $U_{g'}{}^{(\nu)}$ contains the cube with the same center as $U_g{}^{(\nu+1)}$ but twice the side; we set $\rho_{\nu,\nu+1}\ g = g'$. More generally, if $\nu < \mu$, we define

$$\rho_{\nu,\mu}g = \rho_{\nu,\nu+1}\rho_{\nu+1,\nu+2}\cdots\rho_{\mu-1,\mu}g.$$

The maps of I into itself thus defined have the obvious transitivity properties, and the cube $U_g{}^{(\mu)}$ if enlarged $2^{(\mu-\nu)}$ times with the center unchanged will belong to $U_{g'}{}^{(\nu)}$ if $g' = \rho_{\nu,\mu}g$. We shall consistently use the maps $\rho_{\nu,\mu}$ in this section when we consider the coverings $\mathscr{U}^{(\nu)}$ as refinements of each other. The corresponding maps $\rho_{\nu,\mu}^*$ of cochains are then defined as in section 7.3.

Let c be a cochain in $C^p(\mathscr{U}^{(\nu)},\mathscr{A})$, and let φ be a continuous function. We shall then write

$$\|c\|_\varphi^2 = \sum_{|s|=p+1} \int_{U_s{}^{(\nu)}} |c_s|^2\, e^{-\varphi}\, d\lambda$$

where $d\lambda$ denotes the Lebesgue measure in $\mathbf{C}^n$.

Proposition 7.6.1. *For every plurisubharmonic φ in $\mathbf{C}^n$ and every cochain $c \in C^p(\mathscr{U}^{(\nu)},\mathscr{A})$ $(p > 0)$ with $\delta c = 0$ and $\|c\|_\varphi < \infty$, one can find a cochain $c' \in C^{p-1}(\mathscr{U}^{(\nu+p-1)},\mathscr{A})$ so that $\delta c' = \rho_{\nu,\nu+p-1}^* c$ and*

$$\|c'\|_\psi \le K\|c\|_\varphi. \tag{7.6.1}$$

Here K is a constant independent of φ and c, and ψ is defined by

$$\psi(z) = \varphi(z) + 2\log(1 + |z|^2).$$

The main tool in the proof is Theorem 4.4.2, but in addition we need the following local existence theorem for the $\bar{\partial}$ operator.

Lemma 7.6.2. *Let Ω be a domain of holomorphy and let Ω' be a relatively compact subset. For every plurisubharmonic φ in $\mathbf{C}^n$ and every*

$f \in L^2_{(0,q+1)}(\Omega)$ with $\bar{\partial} f = 0$, one can find $u \in L^2_{(0,q)}(\Omega,\mathrm{loc})$ with $\bar{\partial} u = f$ and

$$\int_{\Omega'} |u|^2 e^{-\varphi} \, d\lambda \leq K \int_{\Omega} |f|^2 e^{-\varphi} \, d\lambda,$$

where K is independent of u and of φ.

Proof. Choose functions ψ and $\chi(p)$ as in the proof of Theorem 4.2.2 so that $\chi(p) - \psi$ is bounded from below. Since (4.2.10) is valid with φ replaced by $\varphi + \chi(p)$, it follows from Lemma 4.1.1 (see (4.1.3)) that the equation $\bar{\partial} u = f$ has a solution with

$$\|u\|_{\varphi + \chi(p) - 2\psi} \leq \|f\|_{\varphi + \chi(p) - \psi},$$

if $\varphi \in C^2$. The right-hand side can be estimated by a constant times $\|f\|_{\varphi}$, which proves the lemma when $\varphi \in C^2$. The extension to general φ follows as in the proof of Theorem 4.4.2.

Remark. It is proved in Hörmander [1] that the lemma is valid with $\Omega' = \Omega$ if Ω is bounded, but the proof is more complicated in that case. This result would eliminate the need to pass to a refinement in Proposition 7.6.1 (see section 2.4 in Hörmander [1]). However, we have to refine the covering later on anyway to obtain bounds for the lifting of cochains, so very little would be gained by proving a stronger form of the lemma.

Proof of Proposition 7.6.1. In order to copy the proof of Theorem 7.4.1, we introduce the space $C^p(\mathscr{U}^{(\nu)}, \mathscr{L}_q, \varphi)$ of all alternating cochains $c = \{c_s\}$, $s \in I^{p+1}$, where $c_s \in L^2_{(0,q)}(U_s^{(\nu)}, \varphi)$, $\bar{\partial} c_s = 0$ and

$$\|c\|_{\varphi}^2 = \sum_{|s| = p+1} \int_{U_s^{(\nu)}} |c_s|^2 e^{-\varphi} \, d\lambda < \infty.$$

We wish to prove that if $\delta c = 0$ $(p > 0)$, then one can find $c' \in C^{p-1}(\mathscr{U}^{(\nu+p-1)}, \mathscr{L}_q, \psi)$ so that $\delta c' = \rho^*_{\nu,\nu+p-1} c$ and (7.6.1) holds. For $q = 0$, this assertion is precisely Proposition 7.6.1. We shall prove it assuming, if $p > 1$, that it has already been proved for smaller values of p and all q. As in the proof of Theorem 7.4.1, the success of this induction depends on the fact that we have an arbitrary q.

It is of course no restriction to assume that $\nu = 0$. First we have to reconsider Proposition 7.3.3. Thus choose a non-negative function $\chi \in C_0^\infty(U_0^{(0)})$ such that $\Sigma_g \chi(z - g) = 1$. Such a function can be constructed by taking a function $\chi_0 \in C_0^\infty(U_0^{(0)})$ so that $\chi_0 \geq 0$ with strict inequality in the concentric cube with half the side. Then $\chi(z) = \chi_0(z)/\Sigma \chi_0(z - g)$ has the required properties. Now set

$$b_s = \sum \chi(z - g) c_{g,s}, \qquad s \in I^p.$$

Since the functions $\chi(z - g)$ form a partition of unity subordinate to the covering $\mathscr{U}^{(0)}$, the proof of Proposition 7.3.3 gives that $\delta b = c$. From Cauchy's inequality, we obtain

$$|b_s|^2 \leq \sum \chi(z - g)|c_{g,s}|^2,$$

and if again we use the fact that $\Sigma\chi(z - g) = 1$,

$$\|b\|_\varphi^2 \leq \|c\|_\varphi^2.$$

Let $\bar\partial b$ be the cochain $\in C^{p-1}(\mathscr{U}^{(0)}, \mathscr{L}_{q+1}, \varphi)$ defined by

$$(\bar\partial b)_s = \bar\partial b_s = \sum \bar\partial\chi(z - g) \wedge c_{g,s}.$$

Since no point is in the support of more than 2^n of the functions $\chi(z - g)$, we obtain with a constant K

$$\|\bar\partial b\|_\varphi \leq K\|c\|_\varphi.$$

Now $\delta\bar\partial b = \bar\partial\delta b = \bar\partial c = 0$. If $p > 1$, we can therefore by the inductive hypothesis find a cochain $b' \in C^{p-2}(\mathscr{U}^{(p-2)}, \mathscr{L}_{q+1}, \psi)$ such that $\delta b' = \rho_{0,p-2}^{\ *}\bar\partial b$ and for some constant K_1

$$\|b'\|_\psi \leq K_1\|\bar\partial b\|_\varphi \leq KK_1\|c\|_\varphi.$$

Since $\bar\partial b_s' = 0$ for every $s \in I^{p-1}$ and ψ is plurisubharmonic, we can by Lemma 7.6.2 for every $s \in I^{p-1}$ choose $b_s'' \in L^2_{(0,q)}(U_s^{(p-1)}, \psi)$ so that $\bar\partial b_s'' = b_{s'}'$ in $U_s^{(p-1)}$ if $s' = \rho_{p-2,p-1}s$, and with a constant K_2,

$$\int_{U_s^{(p-1)}} |b_s''|^2 e^{-\psi}\, d\lambda \leq K_2 \int_{U_{s'}^{(p-2)}} |b_{s'}'|^2 e^{-\psi}\, d\lambda.$$

Indeed, modulo translations there are only a finite number of sets $U_s^{(p-1)}$, and if $U_s^{(p-1)}$ is enlarged in the ratio 2 to 1 by a homothetic transformation from its center, then it belongs to $U_{s'}^{(p-2)}$. Now set

$$c' = \rho_{0,p-1}^{\ *}b - \delta b''.$$

Then $\delta c' = \rho_{0,p-1}^{\ *}\delta b = \rho_{0,p-1}^{\ *}c$, and

$$\begin{aligned}\bar\partial c' &= \rho_{0,p-1}^{\ *}\bar\partial b - \delta\bar\partial b'' = \rho_{0,p-1}^{\ *}\bar\partial b - \delta\rho_{p-2,p-1}^{\ *}b' \\ &= \rho_{0,p-1}^{\ *}\bar\partial b - \rho_{p-2,p-1}^{\ *}\rho_{0,p-2}^{\ *}\bar\partial b = 0.\end{aligned}$$

Summing up the estimates for b, b', and b'' given above, we obtain (7.6.1).

It remains to consider the case $p = 1$. The fact that $\delta\bar\partial b = 0$ then means that $\bar\partial b$ defines uniquely a form f of type $(0,q + 1)$ in $\mathbf{C}^n$ with

$\bar{\partial} f = 0$ and

$$\int |f|^2 e^{-\varphi} \, d\lambda \leq \|\bar{\partial} b\|_{\varphi}^2 \leq K^2 \|c\|_{\varphi}^2.$$

By Theorem 4.4.2 we can find a form $u \in L^2_{(0,q)}(\mathbf{C}^n, \psi)$ so that $\bar{\partial} u = f$ and

$$\int |u|^2 e^{-\psi} \, d\lambda \leq \int |f|^2 e^{-\varphi} \, d\lambda.$$

Setting $c_s' = b_s - u$, we thus obtain a cochain with the required properties.

Let P_{jk} ($j = 1, \cdots, p$; $k = 1, \cdots, q$) be polynomials in z, and consider the sheaf homomorphism

$$P: \mathscr{A}^q \to \mathscr{A}^p \tag{7.6.2}$$

defined by mapping $(f_1, \cdots, f_q) \in \mathscr{A}^q$ to $\{\Sigma P_{jk} f_k\}_{j=1}^p$. We wish to give bounds for the cohomology with values in the kernel or the range, both of which are of course coherent analytic sheaves. This will be done by essentially repeating the proof of Theorem 7.4.3, but first we must reconsider all of its local ingredients. The first is the Oka theorem in this special case.

Lemma 7.6.3. *The kernel $\mathscr{R}_P$ of the homomorphism* (7.6.2) *is generated by the germs of all q-tuples $Q = (Q_1, \cdots, Q_q)$ with polynomial components such that*

$$\sum_1^q P_{jk} Q_k = 0, \qquad j = 1, \cdots, p; \tag{7.6.3}$$

in fact, by a finite number of such Q.

Proof. Since the polynomial ring is Noetherian (see Zariski and Samuel [1], p. 201), the module of all Q with polynomial components satisfying (7.6.3) is finitely generated over the polynomial ring so that, if all polynomial solutions of (7.6.3) are generators for $\mathscr{R}_P$, then one can find a finite number of generators. The proof of the lemma can be obtained easily from general facts concerning local rings—for example, the Artin–Rees lemma. However, it is enough for us to note that the proof of the Oka theorem (Theorem 6.4.1) can be applied without change. The detailed verification of this may be supplied by the reader.

Let $(Q_{1l}, \cdots, Q_{ql})$, $l = 1, \cdots, r$, be generators with polynomial components for the sheaf $\mathscr{R}_P$. From Theorem 7.2.9 we then obtain

Lemma 7.6.4. *If Ω is a domain of holomorphy and if $f_k \in A(\Omega)$, $k = 1, \cdots, q$, satisfy the equations $\Sigma_1^q P_{jk} f_k = 0$, $j = 1, \cdots, q$, then one can find $g_l \in A(\Omega)$, $l = 1, \cdots, r$, so that*

$$f_k = \sum_1^r Q_{kl} g_l, \qquad k = 1, \cdots, q. \tag{7.6.4}$$

Conversely, (7.6.4) of course implies that $\Sigma P_{jk} f_k = 0$. We have thus found Q so that the range of Q is identical to the kernel of P. We can therefore concentrate now on studying the range of a homomorphism of the form (7.6.2). Our next purpose is to give a quantitative version of Theorem 6.3.5 in this case.

Proposition 7.6.5. *Let $(P_{jk})(j = 1, \cdots, p, k = 1, \cdots, q)$ be a matrix with polynomial entries and let Ω be a neighborhood of 0. Then there exists a neighborhood Ω' of 0 such that for every $u \in A(\Omega + z)^q$ one can find $v \in A(\Omega' + z)^q$ with $Pv = Pu$, so that with constants C and N which are independent of u and of $z \in \mathbf{C}^n$*

$$\sup_{\Omega' + z} |v| \leq C(1 + |z|)^N \sup_{\Omega + z} |Pu|. \tag{7.6.5}$$

Here $\Omega + z$ denotes $\{\zeta + z ; \zeta \in \Omega\}$ and $|v|^2 = \Sigma_1{}^q |v_j|^2$.

Proof. The first part of the proof of Theorem 6.3.5 can be repeated without any change. In fact, assume that $p > 1$ and that the proposition were already proved for systems involving a smaller number p of equations. In particular, we can then consider the equation

$$\sum_1^q P_{1k} v_k{}^1 = \sum_1^q P_{1k} u_k$$

and conclude that it has a solution v^1 in $\Omega' + z$ such that

$$\sup_{\Omega' + z} |v^1| \leq C(1 + |z|)^N \sup_{\Omega + z} |Pu|.$$

Now set $v = v^1 + w$. We have to find w so that $Pw = P(u - v^1)$ and w can be estimated. Now we can apply Lemma 7.6.4 to the system of equations $\Sigma P_{1k} w_k = 0$. If $(Q_{1l}, \cdots, Q_{ql})$, $l = 1, \cdots, r$, are generators with polynomial coefficients of these relations and if Ω' is a domain of holomorphy, we can write

$$u - v^1 = Qf,$$

where $f = (f_1, \cdots, f_r) \in A(\Omega')^r$. Now we want to find w of the form $w = Qg$. The equation $Pw = P(u - v^1)$ then becomes $PQg = PQf$, and of these p equations the first is satisfied automatically in view of the definition of Q. Hence we have only $p - 1$ equations, so by the inductive hypothesis we can for a suitable neighborhood Ω'' of 0 find g so that with some constants C' and N'

$$\sup_{\Omega''+z} |g| \le C'(1 + |z|)^{N'} \sup_{\Omega'+z} |PQf| = C'(1 + |z|)^{N'} \sup_{\Omega'+z} |Pu - Pv^1|.$$

If m is an upper bound for the degrees of the polynomials P_{jk} and Q_{kl}, we obtain with $v = v^1 + Qg$

$$\sup_{\Omega''+z} |v| \le C''(1 + |z|)^{N+N'+2m} \sup_{\Omega+z} |Pu|,$$

and since $Pv = Pv^1 + PQg = Pv^1 + PQf = Pv^1 + P(u - v^1) = Pu$, this proves the assertion.

It remains, however, to show that the proposition is true when $p = 1$ provided that it is true for a smaller number of variables and an arbitrary p. To do that requires a more detailed study of the properties of polynomials and also another look at the Weierstrass preparation theorem. The end of the proof of Proposition 7.6.5 must be postponed until we have made these preparations.

Shrinking Ω and making a linear change of variables we may of course always assume that Ω is the unit ball.

Lemma 7.6.6. *Given an integer $m > 0$, one can find a finite subset Θ of the unit ball $\Omega = \{\vartheta; \vartheta \in \mathbf{C}^n, |\vartheta| < 1\}$, so that for all polynomials p of degree $\le m$ we have*

$$\sup_{|z|<1} |p(z)| \le C \sup_{\vartheta\in\Theta} \inf_{|w|=1} |p(w\vartheta)|, \tag{7.6.6}$$

where C only depends on n and m.

This is precisely Lemma 3.1.6 in Hörmander [2] and we refer the reader to the proof given there. We also recall that if $p(z) = \Sigma_{|\alpha|\le m} a_\alpha z^\alpha$, then

$$C_1 \sum |a_\alpha| \le \sup_{|\vartheta|<1} |p(z)| \le C_2 \sum |a_\alpha| \tag{7.6.7}$$

for some positive constants depending only on n and m. Indeed, all norms in a finite dimensional vector space are equivalent.

We shall now consider the Weierstrass preparation theorem when the maximum in the right-hand side of (7.6.6) is attained for the vector $\vartheta = (0, \cdots, 0, r)$. We use the notation

$$\tilde{p}(z) = \sup_{|\zeta|<1} |p(z + \zeta)|.$$

It is important to note that $\tilde{p}(z)$ is bounded from below in $\mathbf{C}^n$ if p is not identically 0. Indeed, if $a_\alpha z^\alpha$ is a non-zero term of highest degree in p, it follows from (7.6.7) that $\tilde{p}(z) \geq C_1|a_\alpha|$.

Lemma 7.6.7. *Let p be a polynomial of degree $\leq m$, let $0 < r < 1$ and assume that*

$$\tilde{p}(0) \leq C \inf_{|z_n|=r} |p(0,z_n)|, \tag{7.6.8}$$

where C is the constant in (7.6.6). Then one can find a number r' with $0 < r' < 1$, depending only on r, n, and m, so that the polydisc

$$\Delta = \{z\,;|z_j| < r',\ \ j < n, |z_n| < r\}$$

is contained in the unit ball, and every bounded $f \in A(\Delta)$ can be written

$$f = pg + h,$$

where g, $h \in A(\Delta)$, h is a polynomial in z_n of degree less than the number m^+ of zeros of $p(0,z_n)$ with $|z_n| < r$, and

$$\sup_\Delta |h| \leq C' \sup_\Delta |f|, \qquad \sup_\Delta |g| \leq C' \sup_\Delta |f|/\tilde{p}(0), \tag{7.6.9}$$

where C' is a constant depending only on n and m.

Proof. It is no restriction to assume that $\tilde{p}(0) = 1$. By (7.6.7) this implies that we have uniform bounds for the coefficients of p. Set

$$p(z) = p_0(z_n) + p_1(z)$$

where $p_0(z_n) = p(0,z_n)$ and therefore $p_1(z',0) = 0$. Then we have

$$|p_1(z)| < Kr' \quad \text{in } \Delta \text{ if } \Delta \subset \Omega, \tag{7.6.10}$$

where K is a constant. (In the whole proof we require all constants to depend only on n and m.) Furthermore, since $|p_0(z_n)| \geq 1/C$ when $|z_n| = r$ and we have bounds for the coefficients of p_0, we can find a constant $\varepsilon > 0$ so that

$$|p_0(z_n)| \geq 1/2C \quad \text{when } ||z_n| - r| < \varepsilon.$$

In particular, p_0 has no zeros in this annulus. Now write

$$p_0 = p_0{}^+ p_0{}^-,$$

where the zeros of $p_0{}^+$ lie in the circle $|z_n| < r$ and those of $p_0{}^-$ outside the circle, and the leading coefficient of $p_0{}^+$ is one. Then

$$|p_0^+(z_n)| \geq (|z_n| - r + \varepsilon)^{m^+} \text{ when } |z_n| > r - \varepsilon.$$

Furthermore, we have uniform bounds for the coefficients of p_0^+.

Now we can write every bounded $f \in A(\Delta)$ in the form

$$f = p_0 g + h,$$

where h is a polynomial in z_n of degree $< m^+$ and with a constant K'

$$\sup_\Delta |g| \leq K' \sup_\Delta |f|, \quad \sup_\Delta |h| \leq K' \sup_\Delta |f|. \tag{7.6.11}$$

In fact, we only have to set when $z \in \Delta$

$$g(z) = \frac{1}{2\pi i p_0^-(z_n)} \int \frac{f(z',\tau)}{p_0^+(\tau)(\tau - z_n)} d\tau, \; h(z) = \frac{1}{2\pi i} \int \frac{f(z',\tau)}{p_0^+(\tau)} \frac{p_0^+(\tau) - p_0^+(z_n)}{\tau - z_n} d\tau,$$

where the integration is taken over a circle $|\tau| = \rho$ with $|z_n| < \rho$ and $r - \varepsilon < \rho < r$. The notation z' stands for $(z_1, \cdots, z_{n-1})$. Since $(p_0^+(\tau) - p_0^+(z_n))/(\tau - z_n)$ is a polynomial in z_n, τ, it is clear that h is a polynomial in z_n and that we have the desired bound for h. Hence we obtain a bound for $|p_0 g|$ and therefore

$$|g| \leq K' \sup_\Delta |f|$$

in Δ when $r - \varepsilon < |z_n| < r$. By the maximum principle this gives (7.6.11).

Now we can prove the lemma by the same iterative method that we used in the proof of Theorem 6.1.1. Thus we set $g_0 = 0$ and define $g_k, h_k \in A(\Delta)$ for $k > 0$ by the recursion formula

$$f = p_0 g_k + p_1 g_{k-1} + h_k$$

and the condition that h_k be a polynomial in z_n of degree $< m_+$ for every k. Then g_k and h_k are bounded analytic functions in Δ for every k, and since

$$-p_1(g_k - g_{k-1}) = p_0(g_{k+1} - g_k) + h_{k+1} - h_k,$$

it follows from (7.6.10) and (7.6.11) that

$$\sup_\Delta |g_{k+1} - g_k| \leq KK'r' \sup_\Delta |g_k - g_{k-1}|.$$

If $r' \leq 1/2KK'$, this implies the existence of $g = \lim_{k\to\infty} g_k = \Sigma_0^\infty (g_{k+1} - g_k)$ and that

$$\sup_\Delta |g| \leq 2 \sup_\Delta |g_1| \leq 2K' \sup_\Delta |f|.$$

The existence of $h = \lim h_k$ follows from the recursion formula which also shows that $f = pg + h$. The asserted bound for h now follows immediately.

From Lemma 7.6.7 we obtain an analogue of Corollary 6.1.2 by taking $f(z) = z_n^{m^+}$:

Lemma 7.6.8. *The polydisc Δ in Lemma 7.6.7 can be chosen so that every p of degree $\leq m$ satisfying (7.6.8) can be factored as $p = p^+p^-$, where*

(i) *p^+ and p^- are polynomials in z_n which are analytic and bounded in Δ,*

(ii) *p^+ has leading coefficient 1 and all its zeros in Δ when $|z_j| < r'$, $j < n$,*

(iii) *$\tilde{p}(0)/C' \leq \inf_\Delta |p^-| \leq \sup_\Delta |p^-| \leq C''\tilde{p}(0)$, where C' and C'' are constants depending only on n and m.*

Proof. Since all coefficients of p are $O(\tilde{p}(0))$, we can choose r' so small that for some $\varepsilon > 0$ we have

$$p(z) \neq 0 \quad \text{if } |z_j| < r' \text{ when } j < n, \text{ and } ||z_n| - r| < \varepsilon.$$

Then we conclude that $p(z',z_n)$ has precisely m^+ zeros with $|z_n| \leq r - \varepsilon$ if $|z_j| < r'$, $j < n$. Now Lemma 7.6.7 implies that we can find $g,h \in A(\Delta)$ so that $z_n^{m^+} = pg - h$ and h is a polynomial in z_n of degree $< m^+$. The polynomial $p^+ = z_n^{m^+} + h$ must therefore have all its zeros in the disc $|z_n| \leq r - \varepsilon$, so $|p^+(z)| \geq \varepsilon^{m^+}$ when $|z_n| = r$ and $|z_j| < r'$, $j < n$. This proves that $|g| \geq \varepsilon^{m^+}/\tilde{p}(0)$ when $z \in \Delta$. If we set $p^- = 1/g$ the lemma follows in view of (7.6.9).

Repetition of the proof of Lemma 6.1.3 now gives

Lemma 7.6.9. *Let p and Δ be as in Lemma 7.6.8. If $f \in A(\Delta)$ and pf is a polynomial in z_n, then p^-f is a polynomial in z_n.*

Proof. By the algebraic division algorithm we have $pf = p^+g + h$, where g and h are polynomials in z_n and h is of degree $< m^+$. Since h must vanish at all the m^+ zeros of p^+ for a fixed z', it follows that $h = 0$. Hence $p^-f = g$ is a polynomial in z_n.

End of proof of Proposition 7.6.5. It remains to prove that if Ω is the unit ball and $p = 1$, then the proposition is true if it is true in general when the number of variables is smaller. We may assume that $P_1 \not\equiv 0$ and that the degree m of P_1 is at least as large as the degree of P_j for $j = 2, \cdots, q$. This implies that $\tilde{P}_1(z)$ is bounded from below. Now apply

Lemma 7.6.6 to P_1. For every $\vartheta \in \Theta$ let E_ϑ be the set of all $z \in \mathbf{C}^n$ such that

$$\tilde{P}_1(z) \leq C \inf_{|w|=1} |P_1(z + w\vartheta)|.$$

It is sufficient to prove that for every fixed $\vartheta \in \Theta$ the proposition is valid when $z \in E_\vartheta$. By a linear change of variables we may reduce the proof to the case $\vartheta = (0, \cdots, 0, r)$. Put

$$P_1 u_1 + \cdots + P_q u_q = f$$

and choose Δ so that Lemmas 7.6.7 and 7.6.8 can be applied. For every $z \in E_\vartheta$ we can then write

$$f = P_1 g + h,$$

where $g, h \in A(z + \Delta)$, h is a polynomial in ζ_n of degree $< m^+$, the number of zeros in $z + \Delta$ of the equation $P_1(z', \zeta_n) = 0$, and for some constant K

$$\sup_{z+\Delta} |g| + \sup_{z+\Delta} |h| \leq K \sup_{z+\Omega} |f|. \tag{7.6.12}$$

When $j > 1$ we also write $u_j = P_1 U_j + w_j$, where w_j is a polynomial in ζ_n of degree $< m^+$. With $U_1 = u_1 + P_2 U_2 + \cdots + P_q U_q$ we then obtain

$$P_1 U_1 + P_2 w_2 + \cdots + P_q w_q = P_1 g + h.$$

Set $U_1 = g + w_1$. Then we obtain

$$P_1 w_1 + \cdots + P_q w_q = h.$$

Since all terms except possibly the first are polynomials in ζ_n of degree $< m + m^+$, we conclude that $P_1 w_1$ is a polynomial in ζ_n of degree $< m + m^+$. However, this does not imply that w_1 is a polynomial in ζ_n. In order to be able to apply Lemma 7.6.9, we therefore decompose P_1 by applying Lemma 7.6.8 to the polynomial $P_1(z + \zeta)$. It follows that we can write $P_1 = p^+ p^-$, where p^+ and p^- are analytic in $z + \Delta$, both are polynomials in ζ_n, the ratio between p^- and $\tilde{P}_1(z)$ is bounded from above and below in $z + \Delta$ by constants, and the zeros of p^+ and p^- are located as described in the lemma. Then we obtain that the functions $w_j' = p^- w_j$, $j = 1, \cdots, q$, and $h' = p^- h$ are analytic in $z + \Delta$ and polynomials in ζ_n of degree $< 2m$. We have

$$P_1 w_1' + \cdots + P_q w_q' = h' \quad \text{in } z + \Delta$$

and by (7.6.12) we have the estimate

$$\sup_{z+\Delta} |h'| \le K'\tilde{P}_1(z) \sup_{z+\Omega} |f|. \tag{7.6.13}$$

Now we can write

$$w_j'(\zeta) = \sum_0^{2m-1} w_{jk}(\zeta')\zeta_n{}^k, \qquad h'(\zeta) = \sum_0^{2m-1} h_k(\zeta')\zeta_n{}^k$$

where the coefficients are analytic and bounded in $z' + \Delta'$, and $\Delta' = \{\zeta' \in \mathbf{C}^{n-1}; |\zeta_j| < r', j < m\}$. From the one-dimensional case of (7.6.7) and from (7.6.13), it follows that

$$\begin{aligned} \sum_k \sup_{z'+\Delta'} |h_k(\zeta')| &\le K(1 + |z_n|)^{2m-1} \sup_{z+\Delta} |h'(\zeta)| \\ &\le K'\tilde{P}_1(z)(1 + |z_n|)^{2m-1} \sup_{z+\Delta} |f(\zeta)|. \end{aligned} \tag{7.6.14}$$

Now the equation

$$P_1 w_1' + \cdots + P_q w_q' = h'$$

is equivalent to a system of $3m$ equations in the w_{jk} where the right-hand sides are the functions h_k and zeros. By the inductive hypothesis we can therefore find a neighborhood Δ'' of 0 in $\mathbf{C}^{n-1}$ so that there are analytic functions W_{jk} in $z' + \Delta''$ such that the polynomials

$$W_j(\zeta) = \sum_0^{2m-1} W_{jk}(\zeta')\zeta_n{}^k$$

satisfy the equation

$$P_1 W_1 + \cdots + P_q W_q = h'$$

and

$$\sum \sup_{z'+\Delta''} |W_{jk}| \le K(1 + |z'|)^N \sum \sup_{z'+\Delta'} |h_k|. \tag{7.6.15}$$

Setting

$$v_1 = g + W_1/p^-, \quad v_j = W_j/p^- \quad \text{when } j > 1,$$

we obtain

$$P_1 v_1 + \cdots + P_q v_q = P_1 g + h'/p^- = P_1 g + h = f$$

in the subset of $z + \Delta$ where $\zeta' \in z' + \Delta''$; and from (7.6.12), (7.6.14), (7.6.15), and the fact that p^- can be bounded from below by a constant times $\tilde{P}_1(z)$, we obtain (7.6.5). The proof is finally complete.

Note that a part of the proof follows that of Lemma 6.4.2. Every component was in fact used already in Chapter VI, and it is only keeping track of estimates that makes the proof heavy.

We now have all the prerequisites for proving a Theorem B with bounds. Let P_{jk}, $j = 1, \cdots, p; k = 1, \cdots, q$, be polynomials and set as above

$$\mathscr{R}_P = \{(f_1, \cdots, f_q) \in \mathscr{A}^q, \sum_1^q P_{jk} f_k = 0,\ j = 1, \cdots, p\}.$$

If φ is a continuous function, we define $C^\sigma(\mathscr{U}^{(\nu)}, \mathscr{R}_P, \varphi)$ as the set of alternating cochains $c = \{c_s\}$, $s \in I^{\sigma+1}$, where $c_s \in \Gamma(U_s^{(\nu)}, \mathscr{R}_P)$ and

$$\|c\|_\varphi^2 = \sum_{|s| = \sigma+1} \int_{U_s^{(\nu)}} |c_s|^2 e^{-\varphi} \, d\lambda < \infty.$$

Here we have set $|f|^2 = |f_1|^2 + \cdots + |f_q|^2$ if $f = (f_1, \cdots, f_q) \in A(\Omega)^q$ for some open set Ω.

Theorem 7.6.10. *Let P and an integer ν be given. Then there exist integers μ and N such that, if φ is plurisubharmonic and for some constant C*

$$|\varphi(z) - \varphi(z')| < C, \qquad |z - z'| < 1, \tag{7.6.16}$$

then one can for every $c \in C^\sigma(\mathscr{U}^{(\nu)}, \mathscr{R}_P, \varphi)$ with $\delta c = 0$, $\sigma > 0$, find $c' \in C^{\sigma-1}(\mathscr{U}^{(\mu)}, \mathscr{R}_P, \varphi_N)$ so that $\delta c' = \rho_{\nu,\mu}^* c$ and for some constant K

$$\|c'\|_{\varphi_N} \le K \|c\|_\varphi. \tag{7.6.17}$$

Here $\varphi_N(z) = \varphi(z) + N \log(1 + |z|^2)$.

Proof. First note that Proposition 7.6.5 is also valid if we replace (7.6.5) by

$$\int_{z+\Omega'} |v|^2 e^{-\varphi(z)} (1 + |z|^2)^{-N} \, d\lambda(z) \le C \int_{z+\Omega} |Pu|^2 e^{-\varphi(z)} \, d\lambda(z). \tag{7.6.5)'$$

In fact, by (7.6.16) the factor $e^{-\varphi}$ is immaterial, and in view of Theorem 2.2.3 the new formulation follows when $\varphi = 0$ if we apply Proposition 7.6.5 with Ω replaced by a neighborhood $\omega \subset\subset \Omega$ of $\overline{\Omega'}$.

We shall prove Theorem 7.6.10 by induction for decreasing σ, noting that it is valid when $\sigma > 2^{2n}$, since there are no non-zero $c \in C^\sigma(\mathscr{U}^{(\nu)}, \mathscr{R}_P, \varphi)$ then. Thus assume that the theorem has been proved for all P when σ is replaced by $\sigma + 1$. If Q is chosen as in Lemma 7.6.4, we can write $c_s = Q d_s$, where $d \in C^\sigma(\mathscr{U}^{(\nu)}, \mathscr{A}^r)$. To obtain control of d_s we must pass to a refinement of the covering so that Proposition 7.6.5 can be used. Note that for $\mu > \nu$ the set $U_s^{(\mu)}$ enlarged in the ratio $2^{\mu-\nu}$ with the center kept fixed belongs to $U_{s'}^{(\nu)}$ if $s' = \rho_{\nu,\mu}s$. By the comments on Proposition 7.6.5 made at the beginning of the proof, we can therefore if μ is large choose $d_s' \in A(U_s^{(\mu)})^r$ so that $Qd_s' = Qd_{s'} = c_{s'}$ in $U_s^{(\mu)}$ and

$$\int_{U_s^{(\mu)}} |d_s'|^2(1 + |z|^2)^{-N} e^{-\varphi(z)}\, d\lambda(z) \le C \int_{U_{s'}^{(\nu)}} |c_{s'}|^2 e^{-\varphi(z)}\, d\lambda(z).$$

Summing up, we have

$$\|d'\|_{\varphi_N} \le C\|c\|_\varphi$$

and $\rho_{\nu,\mu}^*c = Qd'$. (By C we denote different constants in different estimates.) Since $\delta c = 0$, it follows that $\delta Qd' = Q\delta d' = 0$. Thus $\delta d' = d'' \in C^{\sigma+1}(\mathscr{U}^{(\mu)}, \mathscr{R}_Q, \varphi_N)$, and since $\delta d'' = 0$ and φ_N is plurisubharmonic, it follows by the inductive hypothesis that for suitable N' and $\mu' > \mu$ we can find $d''' \in C^\sigma(\mathscr{U}^{(\mu')}, \mathscr{R}_Q, \varphi_{N'})$ so that $\delta d''' = \rho_{\mu,\mu'}^*d''$ and

$$\|d'''\|_{\varphi_{N'}} \le C\|d''\|_{\varphi_N}.$$

Setting $\gamma = \rho_{\mu,\mu'}^*d' - d''' \in C^\sigma(\mathscr{U}^{(\mu')}, \mathscr{A}^r)$, we have $\delta\gamma = \rho_{\mu,\mu'}^*d'' - \delta d''' = 0$ and

$$\|\gamma\|_{\varphi_{N'}} \le C\|c\|_\varphi.$$

Hence Proposition 7.6.1 shows that for some $\mu'' > \mu'$ one can find $\gamma' \in C^{\sigma-1}(\mathscr{U}^{(\mu'')}, \mathscr{A}^r)$ so that $\rho_{\mu',\mu''}^*\gamma = \delta\gamma'$ and for some N''

$$\|\gamma'\|_{\varphi_{N''}} \le C\|\gamma\|_{\varphi_{N'}} \le C'\|c\|_\varphi. \tag{7.6.18}$$

If we set $c' = Q\gamma'$, it follows that

$$\delta c' = Q\delta\gamma' = Q\rho_{\mu',\mu''}^*\gamma = Q\rho_{\mu',\mu''}^*\rho_{\mu,\mu'}^*d' - \rho_{\mu',\mu''}^*Qd'''$$

$$= \rho_{\mu,\mu''}^*Qd' = \rho_{\mu,\mu''}^*\rho_{\nu,\mu}^*c = \rho_{\nu,\mu''}^*c.$$

Since (7.6.18) implies (7.6.17) for a suitable choice of N, the theorem is proved.

In the main applications which we shall now give, the cochains will finally disappear.

Theorem 7.6.11. *Given the system P there is a constant N such that, if φ is a plurisubharmonic function satisfying* (7.6.16), *then for all $u \in A(\mathbf{C}^n)^q$ one can find $v \in A(\mathbf{C}^n)^q$ with $Pv = Pu$ and*

$$\int |v|^2 e^{-\varphi}(1 + |z|^2)^{-N}\, d\lambda(z) \le C \int |Pu|^2 e^{-\varphi}\, d\lambda(z), \tag{7.6.19}$$

where C does not depend on u.

Proof. There is nothing to prove unless

$$\int |Pu|^2 e^{-\varphi} d\lambda(z) < \infty,$$

so we assume that this condition is fulfilled. By application of Proposition 7.6.5, as in the proof of Theorem 7.6.10, we conclude that ν can be chosen so that for every $g \in I$ there exists an element $u_g \in A(U_g^{(\nu)})^q$ such that $Pu_g = Pu$ in $u_g^{(\nu)}$ and for some constants C and N independent of u and g

$$\int_{U_g^{(\nu)}} |u_g|^2 e^{-\varphi}(1 + |z|^2)^{-N} d\lambda(z) \leq C \int_{U_{g'}^{(0)}} |Pu|^2 e^{-\varphi} d\lambda(z) \tag{7.6.20}$$

where $g' = \rho_{0,\nu}g$.

There is no reason why the u_g should agree in the overlaps of the cubes $U_g^{(\nu)}$, so we have to consider the differences

$$c_{g_1 g_2} = u_{g_1} - u_{g_2}.$$

This defines a cocycle $c \in C^1(\mathscr{U}^{(\nu)}, \mathscr{R}_P, \varphi)$, and by (7.6.20) we obtain

$$\|c\|^2_{\varphi_N} \leq C \int |Pu|^2 e^{-\varphi} d\lambda(z). \tag{7.6.21}$$

(C denotes different constants in different estimates.) Now Theorem 7.6.10 asserts that for some $\mu > \nu$ and $N' > N$ there exists a cochain $c' \in C^0(\mathscr{U}^{(\mu)}, \mathscr{R}_P, \varphi_{N'})$ such that $\delta c' = \rho^*_{\nu,\mu} c$ and

$$\|c'\|_{\varphi_{N'}} \leq C \|c\|_{\varphi_N}. \tag{7.6.22}$$

This means that if we set

$$v = u_{\rho_{\nu,\mu} g} + c_g' \text{ in } U_g^{(\mu)},$$

we define uniquely an element $v \in A(\mathbf{C}^n)^q$. Since $Pc_g' = 0$, it follows that $Pv = Pu$, and from the estimates (7.6.20), (7.6.21) and (7.6.22) we obtain (7.6.19) with N replaced by N'. The proof is complete.

Combination of Lemma 7.6.4 with Theorem 7.6.11 also gives

Corollary 7.6.12. *Given P there is an integer N such that if φ is plurisubharmonic and satisfies (7.6.16) and if $f \in A(\mathbf{C}^n)^q$ satisfies the equations $Pf = 0$, then $f = Qg$ with*

$$\int |g|^2 e^{-\varphi}(1 + |z|^2)^{-N} d\lambda(z) \leq C \int |f|^2 e^{-\varphi} d\lambda(z)$$

for some C independent of f. Here Q has the same meaning as in Lemma 7.6.4.

Note that Theorem 4.4.4 shows that Theorem 7.6.11 and Corollary 7.6.12 are not vacuous for any φ.

We shall now prove existence theorems for systems of partial differential equations with constant coefficients by using Theorem 7.6.11 and Corollary 7.6.12 in combination with Fourier transforms and duality.

Let P_{jk}, $j = 1, \cdots, J, k = 1, \cdots, K$ be polynomials in n variables with complex coefficients and consider the system of differential equations

$$\sum_{1}^{K} P_{jk}(D)u_k = f_j, \qquad j = 1, \cdots, J, \tag{7.6.23}$$

where $D = -i\partial/\partial x$, in an open set $\Omega \subset \mathbf{R}^n$. Consider the set of all J-tuples $Q = (Q_1, \cdots, Q_J)$ of polynomials such that

$$\sum_{1}^{J} Q_j P_{jk} = 0, \qquad k = 1, \cdots, K.$$

This is a finitely generated module over the polynomial ring. Let $(Q_{11}, \cdots, Q_{1J}), \cdots (Q_{I1}, \cdots, Q_{IJ})$ be a set of generators. If we write (7.6.23) in the form $P(D)u = f$ and interpret $Q(D)$ similarly, we have $Q(D)P(D) = 0$, so a necessary condition for the existence of a solution of (7.6.23) is that $Q(D)f = 0$. Conversely, we shall prove

Theorem 7.6.13. *Let Ω be an open convex set in $\mathbf{R}^n$. Then there is an integer N such that the system of equations $P(D)u = f$ has a solution $u \in C^{\nu}(\Omega)^K$ for all $f \in C^{\nu+N}(\Omega)^J$ such that $Q(D)f = 0$.*

We shall also prove an approximation theorem analogous to the Runge theorem. By an exponential solution of the equation $P(D)u = 0$ we then mean a solution of the form $u(x) = e^{i\langle x,\zeta\rangle}u_0(x)$, where the components of u_0 are polynomials.

Theorem 7.6.14. *Let Ω be an open convex set in $\mathbf{R}^n$. Then the closure in $C^{\nu}(\Omega)^K$ of the linear combinations of exponential solutions of the system $P(D)u = 0$ consists of all solutions in $C^{\nu}(\Omega)^K$ of this system.*

We shall start by proving Theorem 7.6.14 since it is needed in the proof of Theorem 7.6.13 just as the Runge theorem was required in proving for example Theorem 1.4.4.

Proof of Theorem 7.6.14. Let L be a continuous linear form on $C^{\nu}(\Omega)^K$ which is orthogonal to all exponential solutions. The restriction of L to $C^{\infty}(\Omega)^K$ is then a K-tuple $f = (f_1, \cdots, f_K)$ of distributions in $\mathscr{E}'(\Omega)$. Let M be a compact convex subset of Ω containing the support of f, and let H be the supporting function of M,

$$H(\xi) = \sup_{x \in M}\langle x,\xi\rangle.$$

By the Paley–Wiener theorem (see, e.g., Hörmander [2], section 1.7), we have for some integer N

$$\int |\hat{f}(\zeta)|^2 \, e^{-2H(\operatorname{Im}\zeta)}(1 + |\zeta|^2)^{-N} \, d\lambda(\zeta) < \infty. \tag{7.6.24}$$

Denote by ${}^tP(D)$ the transpose of the matrix $P(-D)$. If we can prove that

$$f = {}^tP(D)g \tag{7.6.25}$$

for some $g \in \mathscr{E}'(\Omega)^J$, it will follow that

$$L(u) = \langle u, f \rangle = \langle P(D)u, g \rangle$$

for all $u \in C^\infty(\Omega)^K$. Hence $L(u) = 0$ if $P(D)u = 0$ in a neighborhood of $\operatorname{supp} g$. Since L is continuous on $C^\nu(\Omega)^K$, it follows by applying this result to $u * \chi_\varepsilon$, with χ_ε defined as in the proof of Lemma 4.1.4, that $L(u) = 0$ for every $u \in C^\nu(\Omega)^K$ with compact support in Ω such that $Pu = 0$ in a neighborhood of $\operatorname{supp} g$. Hence $L(u) = 0$ for all $u \in C^k(\Omega)$ satisfying the equation $P(D)u = 0$, and the theorem follows from the Hahn–Banach theorem. Thus it only remains to prove the existence of g. (For the argument given so far, see also Hörmander [2], p. 77.)

The equation $f = {}^tP(D)g$ is equivalent to $\hat{f}(\zeta) = {}^tP(\zeta)\hat{g}(\zeta)$, where $\hat{f}$ and $\hat{g}$ denote the Fourier–Laplace transforms. The construction of $\hat{g}$ is made in three steps:

1. For every ζ there is a formal power series G_ζ at ζ such that at ζ

$$\hat{f} = {}^tPG_\zeta.$$

In fact, this is an infinite system of linear equations for the coefficients of G_ζ of the form discussed in Lemma 6.3.7, so it has a solution if it is compatible. Now the equations can be written

$$D^\alpha \hat{f}_k(\zeta) = D^\alpha \Big(\sum_j P_{jk}(-z)G_j(z)\Big)_{z=\zeta},$$

where α is any multi-order and $k = 1, \cdots, K$. The compatibility means that, if q_j are arbitrary polynomials, then

$$\sum_k q_k(D)\Big(\sum_j P_{jk}(-z)G_j(z)\Big)_{z=\zeta} = 0 \quad \text{for all } G_j \Rightarrow \sum q_k(D)\hat{f}_k(\zeta) = 0. \tag{7.6.26}$$

Set

$$u_k(x) = q_k(-x)\, e^{-i\langle x,\zeta\rangle}.$$

Then

$$\sum P_{jk}(D)u_k(x) = \sum P_{jk}(D_x)q_k(D_\zeta)e^{-i\langle x,\zeta\rangle} = \sum q_k(D_\zeta)(P_{jk}(-\zeta)e^{-i\langle x,\zeta\rangle}),$$

so if the left-hand side of (7.6.26) is true we obtain that $Pu = 0$ by taking $G_j(z) = e^{-i\langle x,z\rangle}$ and all other $G_i = 0$. We have by assumption

$$0 = \langle u,f\rangle = \sum \langle q_k(-x)e^{-i\langle x,\zeta\rangle}, f_k\rangle = \sum q_k(D)\hat{f}_k(\zeta).$$

Hence (7.6.26) holds true and the equation $f = {}^tPG_\zeta$ has a solution in formal power series at ζ.

2. By Corollary 6.3.6 we can now find analytic solutions of the equation $f = {}^tPG$ in a neighborhood of any point in $\mathbf{C}^n$, so by Theorem 7.2.9 it follows that there exist entire solutions of this equation.

3. We can now apply Theorem 7.6.11, using (7.6.24) and the fact that $H(\operatorname{Im}\zeta)$ is convex and therefore plurisubharmonic. It follows that there is a J-tuple G of analytic functions such that $f = {}^tPG$ and for a suitable integer N'

$$\int |G(\zeta)|^2 e^{-2H(\operatorname{Im}\zeta)}(1 + |\zeta|^2)^{-N'} d\lambda(\zeta) < \infty.$$

In view of Theorem 2.2.3 and the fact that H is Lipschitz continuous, this implies that

$$|G(\zeta)| \le Ce^{H(\operatorname{Im}\zeta)}(1 + |\zeta|)^{N'}$$

for some constant C. By the Paley–Wiener theorem we therefore have $G = \hat{g}$, where $g \in \mathscr{E}'(M)^J$. The proof is complete.

We remark that the proof also applies to approximation of C^∞ solutions on a compact convex set.

Proof of Theorem 7.6.13. Let ω be a convex relatively compact subset of Ω. We shall first construct a solution of the equation $Pu = f$ in ω. This equation means that if $v \in C_0^\infty(\omega)^J$, then

$$\langle f,v\rangle = \langle Pu,v\rangle = \langle u,{}^tPv\rangle.$$

To construct u we thus have to extend the linear form

$${}^tPv \to \langle f,v\rangle, \qquad v \in C_0^\infty(\omega)^J.$$

A priori it is not even clear that it is uniquely defined.

Let H be the supporting function of ω and set

$$|v|_\nu^2 = \int |\hat{v}(\zeta)|^2 e^{-2H(\operatorname{Im}\zeta)}(1 + |\zeta|^2)^{-\nu} d\lambda(\zeta).$$

If $v \in C_0^\infty(\omega)^J$ and ${}^tP(D)v = w$, we know from Theorem 7.6.11 that there is a J-tuple V of entire functions such that ${}^tP(\zeta)V(\zeta) = \hat{w}(\zeta)$ and

$$\int |V(\zeta)|^2 e^{-2H(\operatorname{Im}\zeta)}(1 + |\zeta|^2)^{-\nu-N} d\lambda(\zeta) \le C|w|_\nu^2, \tag{7.6.27}$$

where C and N are constants. By the Paley–Wiener theorem we have $V = \hat{v}_1$ for some $v_1 \in \mathscr{E}'(\bar{\omega})^J$ of order at most $\nu + N$. Note that ${}^tP(D)v_1 = {}^tP(D)v = w$. Since ${}^tP(\zeta)(V(\zeta) - \hat{v}(\zeta)) = 0$ and the columns of ${}^tQ$ are generators for $\mathscr{R}_{{}^tP}$, we have by Corollary 7.6.12.

$$V(\zeta) - \hat{v}(\zeta) = {}^tQ(\zeta)\Phi(\zeta),$$

where for a certain integer N' depending only on P and on Q

$$\int |\Phi(\zeta)|^2 \, e^{-2H(\operatorname{Im}\zeta)}(1 + |\zeta|^2)^{-\nu-N'} \, d\lambda(\zeta) < \infty.$$

From the Paley–Wiener theorem it follows that $\Phi = \hat{\varphi}$ where the support of φ belongs to $\bar{\omega}$. If $F \in C^\infty(\Omega)^J$, we conclude that

$$\langle F, v_1\rangle - \langle F, v\rangle = \langle F, {}^tQ(D)\varphi\rangle = \langle Q(D)F, \varphi\rangle,$$

and from this it follows as in the beginning of the proof of Theorem 7.6.14 that

$$\langle f, v\rangle = \langle f, v_1\rangle$$

if $f \in C^{\nu+N}(\Omega)^J$ and $Q(D)f = 0$. Choose $f_0 \in C_0^{\nu+N}(\Omega)^J$ so that $f = f_0$ in a neighborhood of $\bar{\omega}$. Then we obtain

$$|\langle f, v\rangle|^2 = |\langle f, v_1\rangle|^2 = |\langle f_0, v_1\rangle|^2 \leq \Big(\int |\hat{f}_0(-\xi)\hat{v}_1(\xi)| \, d\xi\Big)^2$$
$$\leq \int |\hat{v}_1(\xi)|^2(1 + |\xi|^2)^{-\nu-N} \, d\xi \int |\hat{f}_0(-\xi)|^2(1 + |\xi|^2)^{\nu+N} \, d\xi,$$

where the last integral converges since $\hat{f}_0(\xi)\xi^\alpha \in L^2$ when $|\alpha| \leq \nu + N$. From (7.6.27) we thus obtain

$$|\langle f, v\rangle| \leq C|{}^tPv|_\nu.$$

The linear form

$${}^tPv \to \langle f, v\rangle, \quad v \in C_0^{\infty}(\omega)^J,$$

can therefore be extended by the Hahn–Banach theorem to a linear form on the L^2 space in $\mathbf{C}^n$ with respect to the measure

$$e^{-2H(\operatorname{Im}\zeta)}(1 + |\zeta|^2)^{-\nu} \, d\lambda(\zeta).$$

Hence there is a K-tuple U of measurable functions such that

$$\int |U(\zeta)|^2 \, e^{2H(\operatorname{Im}\zeta)}(1 + |\zeta|^2)^{\nu} \, d\lambda(\zeta) < \infty \tag{7.6.28}$$

and

$$\langle f, v\rangle = \int \langle U(\zeta), \ {}^tP(\zeta)\hat{v}(\zeta)\rangle \, d\lambda(\zeta), \qquad v \in C_0^{\infty}(\omega)^J.$$

Transposing the matrix P and introducing the definition of $\hat{v}$, we obtain

$$f(x) = \int P(-\zeta)U(\zeta)e^{-i\langle x,\zeta\rangle}\,d\lambda(\zeta), \qquad x \in \omega, \tag{7.6.29}$$

provided that $\nu - n - 1$ exceeds the degree of P so that the integrals are absolutely convergent. If we set

$$u(x) = \int U(\zeta)e^{-i\langle x,\zeta\rangle}\,d\lambda(\zeta), \qquad x \in \omega,$$

it follows from (7.6.28) that $u \in C^{\nu-n-1}(\omega)$ and by (7.6.29) we have $P(D)u = f$ in ω. If ν is replaced by $\nu + n + 1$, we have thus proved that for every $f \in C^{\nu+N+n+1}(\Omega)$ with $Q(D)f = 0$ the equation $P(D)u = f$ has a solution $u \in C^{\nu}(\omega)$, provided that ν exceeds the degree of P.

To complete the proof we just have to take an increasing sequence of convex open sets $\omega_j \subset\subset \Omega$ with union Ω and determine in each ω_j a solution $u_j \in C^{\nu}(\omega_j)^K$ of the equation $P(D)u_j = f$. Using Theorem 7.6.14 we can make the sequence u_j converge to a solution $u \in C^{\nu}(\Omega)^K$ of $P(D)u = f$. The details of this argument are left to the reader since it has already been repeated several times from section 1.4 on.

It is a simple matter to extend Theorem 7.6.13 to an existence theorem when f is a distribution of finite order in Ω. The theorem remains true even if f is of infinite order, but the proof requires additional considerations then.

Notes. The results of sections 7.1 through 7.4 can all be found in the seminars of Cartan [1], to which we refer for further applications. For example, we have not proved here that the sheaf of an analytic set is coherent. Thus we have not included the result for which the theory was originally intended, namely that analytic sets are defined by global equations. The proofs we give may look rather different superficially from those in Cartan [1], for we have available from the start some special cases of Theorems A and B proved in Chapter V. However, the main points are the same. Thus in the proof of Theorem 7.2.1 (the existence of global sections of a coherent analytic sheaf), we use the induction procedure of Cartan [1], although we rely on the existence theorems for sections of analytic vector bundles proved in section 5.6 instead of the theorem of Cartan [2] concerning invertible holomorphic matrix functions. This is no great simplification, but it eliminates the only nonlinear element in the theory. The Dolbeault isomorphism (Theorem 7.4.1) is due to Dolbeault [1]. The analogue in the case of the exterior differential operator is the de Rham theorem which is proved in section 7.5 with the now standard method which goes back to Weil [2]; the proof of the Dolbeault isomorphism is parallel. The presentation in sections 7.4 and 7.5 follows to a large extent that of Malgrange [1] with the modifications imposed by our different starting point.

Results of the type discussed in section 7.6 were first announced by Ehrenpreis [2]. A more detailed version was given at the conference on harmonic analysis at

Stanford in August 1961. The complete proofs are given in Ehrenpreis [4]. They are based on the classical techniques of Oka and Cartan which do not give bounds as easily as the L^2 estimates we have used here. Thus Theorem 7.6.11 and Corollary 7.6.12 are only given by Ehrenpreis for some functions φ which are particularly important in the applications to differential equations. Since Ehrenpreis' announcement in 1960 most of his statements have been proved by Malgrange [2] and Palamodov [1, 2]. All three authors give much stronger versions of Theorem 7.6.14 where solutions of the differential equation $P(D)u = 0$ are represented as absolutely convergent integrals over exponential solutions. This requires a more detailed local analysis than we have given here, but the passage from local to global results involves nothing beyond the results proved in section 7.6.

BIBLIOGRAPHY

We give here a list of the papers quoted in the text either as a source for the material presented or as a reference for further study. We have not aimed at completeness, and numerous fundamental papers connected with the topics discussed are therefore not included.

Andreotti, A., and Grauert, H. [1] Théorèmes de finitude pour la cohomologie des espaces complexes, *Bull. Soc. Math. France* **90,** 193–259 (1962).

Andreotti, A., and Vesentini, E. [1] Carleman estimates for the Laplace–Beltrami equations on complex manifolds, *Publ. Math. Inst. Hautes Etudes Sci.* **25,** 81–130 (1965).

Arens, R. F., and Calderón, A. P. [1] Analytic functions of several Banach algebra elements, *Ann. Math.* (2) **62,** 204–216 (1955).

Ash, M. E. [1] The basic estimate of the $\bar{\partial}$-Neumann problem in the non-Kählerian case, *Amer. J. Math.* **86,** 247–254 (1964).

Behnke, H., and Stein, K. [1] Konvergente Folgen von Regularitätsbereiche, *Math. Ann.* **116,** 204–216 (1938).

Bishop, E. [1] Mappings of partially analytic spaces, *Amer. J. Math.* **83,** 209–242 (1961).

——— [2] Lecture notes, University of California, Berkeley, Calif., 1963.

Bochner, S. [1] Analytic and meromorphic continuation by means of Green's formula, *Ann. Math.* (2) **44,** 652–673 (1943).

Bochner, S., and Martin, W. T. [1] Functions of several complex variables, Princeton University Press, Princeton, N.J., 1948.

Bombieri, E. [1] Algebraic values of meromorphic maps. Inv. Math. 10, 267–287 (1970).

Bremermann, H. [1] Complex convexity, *Trans. Amer. Math. Soc.* **82,** 17-51 (1956).

——— [2] Über die Äquivalenz der pseudokonvexen Gebiete und der Holomorphiegebiete im Raum von *n* komplexen Veränderlichen, *Math. Ann.* **128,** 63–91 (1954).

Browder, A. [1] Cohomology of maximal ideal spaces, *Bull. Amer. Math. Soc.* **67,** 515–516 (1961).

Cartan, H. [1] Séminaires E.N.S., 1951/52.

——— [2] Sur les matrices holomorphes de *n* variables complexes, *J. Math. Pures Appl.* **19,** 1–26 (1940).

Cartan, H., and Thullen, P. [1] Regularitäts- und Konvergenzbereiche, *Math. Ann.* **106,** 617–647 (1932).

Dolbeault, P. [1] Sur la cohomologie des variétés analytiques complexes, *C. R. Acad. Sci. Paris* **236,** 175–177 (1953).

Ehrenpreis, L. [1] A new proof and an extension of Hartog's theorem, *Bull. Amer. Math. Soc.* **67,** 507–509 (1961).

——— [2] A fundamental principle for systems of linear differential equations with constant coefficients and some of its applications, *Proc. Intern. Symp. on Linear Spaces*, 161–174, Jerusalem, 1961.

——— [3] Mean periodic functions, *Amer. J. Math.* **77,** 293–328 (1955).

——— [4] Fourier analysis in several complex variables. Pure and applied Mathematics XVII. Wiley-Interscience Publ. New York 1970.

Garabedian, P., and Spencer, D. C. [1] Complex boundary problems, *Trans. Amer. Math Soc.* **73,** 223–242 (1952).

Grauert, H. On Levi's problem and the imbedding of real-analytic manifolds, *Ann. Math.* (2) **68,** 460–472 (1958).

Harvey, F. R. and Wells, R. O. [1] Holomorphic approximation and hyperfunction theory on a C^1 totally real submanifold of a complex manifold. Math. Ann. 197, 287–318 (1972).

Helson, H., Kahane, J. P., Katznelson, Y., and Rudin, W. [1] The functions which operate on Fourier transforms, *Acta Math.* **102,** 135–157 (1959).

Hervé, M. [1] Several complex variables. Local theory. Oxford University Press, London, and Tata Institute of Fundamental Research, Bombay, 1963.

Hörmander, L. [1] L^2 estimates and existence theorems for the $\bar{\partial}$ operator, *Acta Math.* **113,** 89–152 (1965).

——— [2] Linear partial differential operators, Springer-Verlag and Academic Press, Inc., New York, 1963.

Hörmander, L. and Wermer, J. [1] Uniform approximation on compact sets in $\mathbf{C}^n$. Math. Scand. 23, 5–21 (1968).

Kallin, E. [1] A non-local function algebra, *Proc. Nat. Acad. Sci.* **49,** 821–824 (1963).

Kiselman, C. [1] On unique supports of analytic functionals, *Arkiv för Matematik* **6** (1965).

Kohn, J. J. [1] Harmonic integrals on strongly pseudo-convex manifolds I, *Ann. Math.* (2) **78,** 206–213 (1963).

——— [2] Harmonic integrals on strongly pseudo-convex manifolds II, *Ann. Math.* **79,** 450–472 (1964).

Kohn, J. J., and Nirenberg, L. [1] Non-coercive boundary problems, *Comm. Pure and Appl. Math.,* **18,** 443–492 (1965).

Kohn, J. J., and Rossi, H. [1] On the extension of holomorphic functions from the boundary of a complex manifold, *Ann. Math.* (2) **81,** 451–472 (1965).

Lelong, P. [1] La convexité et les fonctions analytiques de plusieurs variables complexes, *J. Math. Pures Appl.* **31,** 191–219 (1952).

Lewy, H. [1] On the local character of the solution of an atypical linear differential equation in three variables and a related theorem for regular functions of two complex variables, *Ann. Math.* (2) **64,** 514–522 (1956).

Loomis, L. H. [1] An introduction to abstract harmonic analysis, D. Van Nostrand Co., Inc., Princeton, N.J., 1953.

Malgrange, B. [1] Lectures on the theory of functions of several complex variables, Tata Institute of Fundamental Research, Bombay, 1958.

——— [2] Sur les systèmes différentiels à coefficients constants, *Coll. C.N.R.S.*, 113–122, Paris, 1963.

——— [3] Existence et approximation des solutions des équations aux dérivées partielles et des équations de convolution, *Ann. Inst. Fourier* **6,** 271–354 (1955).

Martineau, A. [1] Sur les fonctionnelles analytiques et la transformation de Fourier-Borel, *J. Analyse Math.* **9,** 1–164 (1963).

Mergelyan, S. N. [1] Uniform approximation of functions of a complex variable, *Usp. Mat. Nauk* (*N.S.*) **7,** No. 2 (48), 31–122 (1952). Also in *Amer. Math. Soc. Transl.* **101**.

Milnor, J. [1] Morse theory, *Ann. Math. Studies*, Princeton University Press, Princeton, N.J., 1959.

Morrey, C. B. [1] The analytic embedding of abstract real analytic manifolds, *Ann. Math.* (2) **68,** 159–201 (1958).

Naimark, M. [1] Normed rings, Groningen, 1959.

Narasimhan, R. [1] Holomorphically complete complex spaces, *Amer. J. Math.* **82,** 917–934 (1961).

Newlander, A., and Nirenberg, L. [1] Complex analytic coordinates in almost complex manifolds, *Ann. Math.* (2) **65,** 391–404 (1957).

Nirenberg, L. [1] Partial differential equations with applications in geometry, Lectures on modern mathematics, Vol. II, John Wiley and Sons, New York, 1964, pp. 1–41.

Norguet, F. [1] Sur les domaines d'holomorphie des fonctions uniformes de plusieurs variables complexes, *Bull. Soc. Math. France* **82,** 137–159 (1954).

Oka, K. [1] Sur les fonctions de plusieurs variables. I. Domaines convexes par rapport aux fonctions rationnelles, *J. Sc. Hiroshima Univ.* **6,** 245–255 (1936).

——— [2] Sur les fonctions de plusieurs variables. II. Domaines d'holomorphie. *J. Sc. Hiroshima Univ.* **7,** 115–130 (1937).

——— [3] Sur les fonctions de plusieurs variables. III. Deuxième problème de Cousin, *J. Sc. Hiroshima Univ.* **9,** 7–19. (1939).

——— [4] Sur les fonctions de plusieurs variables. VII. Sur quelques notions arithmétiques, *Bull. Soc. Math. France* **78,** 1–27 (1950).

——— [5] Sur les fonctions de plusieurs variables. IX. Domaines finis sans point critique intérieur, *Jap. J. Math.* **23,** 97–155 (1953).

Osgood, W. F. [1] Lehrbuch der Funktionentheorie II_1, Leipzig, 1924.

Palamodov, V. P. [1] The general theorems on the system of linear equations with constant coefficients. Outlines of the joint Soviet-American symposium on partial differential equations, August 1963, pp. 206–213.

——— [2] Linear differential operators with constant coefficients. Grundl. d. Math. Wiss. 168. Springer-Verlag, New York, Heidelberg, Berlin 1970.

Pólya, G. [1] Untersuchungen über Lücken und Singularitäten von Potenzreihen, *Math. Z.* **19,** 549–640 (1929).

Radó, T. [1] Subharmonic functions, Berlin, 1937.

de Rham, G. [1] Variétés différentiables, Paris, 1955.

Rossi, H. [1] The local maximum modulus principle. *Ann. Math.* (2) **72,** 1–11 (1960).

Royden, H. L. [1] Function algebras, *Bull. Amer. Math. Soc.* **69,** 281–298 (1963).

Schwartz, L. Théorie des distributions I, Paris, France, 1950.

Serre, J. P. [1] Faisceaux algébriques cohérents, *Ann. Math.* (2), **61,** 197–278 (1955).

——— [2] Une propriété topologique des domaines de Runge, *Proc. Amer. Math. Soc.* **6,** 133–134 (1955).

——— [3] Quelques problèmes globaux relatifs aux variétés de Stein. Coll. sur les fonctions de plusieurs variables, Bruxelles 1953, 57–68.

Shilov, G. E. [1] On decomposition of a commutative normed ring in a direct sum of ideals, Mat. Sb. (N.S.) 32(74) (1953), 353–364 (Russian). Also in *Amer. Math. Soc. Transl.* (2) **1,** 37–48 (1955).

Skoda, H. [1] Sous-ensembles analytiques d'ordre fini ou infini dans $\mathbf{C}^n$. Bull. Soc. Math. France 100, 353–408 (1972).

——— [2] Application des techniques L^2 à la théorie des idéaux d'une algèbre de fonctions holomorphes avec poids. Ann. Ec. Norm. Sup. 5(4), 545–579 (1972).

Stein, K. [1] Analytische Funktionen mehrerer komplexer Veränderlichen und das zweite Cousin'sche Problem, *Math. Ann.* **123,** 201–222 (1951).

Weil, A. [1] Variétés Kähleriennes, Paris, 1957.

——— [2] Sur les théorèmes de de Rham, *Comm. Math. Helv.*, **26,** 119–145 (1952).

——— [3] L'intégrale de Cauchy et les fonctions de plusieurs variables, *Math. Ann.* **111,** 178–182 (1935).

Wermer, J. [1] Seminar über Funktionenalgebren, Springer lecture notes in mathematics 1 (1964).

Zariski, O., and Samuel, P. [1] Commutative algebra, Vol. I, D. Van Nostrand, Co., Inc., Princeton, N.J., 1958.

INDEX